HUMAN CELL CULTURE
Volume V: Primary Mesenchymal Cells

Human Cell Culture

Volume 5

The titles published in this series are listed at the end of this volume.

Human Cell Culture

Volume V

Primary Mesenchymal Cells

edited by

Manfred R. Koller

Oncosis, San Diego, California, USA

Bernhard O. Palsson

Department of Bioengineering, University of California San Diego, USA

and

John R.W. Masters

University College London, Institute of Urology and Nephrology, London, UK

SPRINGER-SCIENCE+BUSINESS MEDIA, B.V.

Library of Congress Cataloging-in-Publication data is available.

ISBN 978-90-481-5633-7 ISBN 978-0-306-46870-4 (eBook)
DOI 10.1007/978-0-306-46870-4

Printed on acid-free paper

Originally published by Kluwer Academic Publishers in 2001
Softcover reprint of the hardcover 1st edition 2001

Contents

Introduction

The human body contains many specialized tissues that are capable of fulfilling an incredible variety of functions necessary for our survival. This volume in the *Human Cell Culture Series* focuses on mesenchymal tissues and cells. The *in vitro* study of mesenchymal cells is perhaps the oldest form of human cell culture, beginning with the culturing of fibroblasts. Fibroblasts have long been generically described in the literature, arising from many tissue types upon *in vitro* cell culture. However, recent studies, many enabled by new molecular biology techniques, have shown considerable diversity in fibroblast type and function, as described within this volume.

Mesenchymal tissue types that are described within include bone, cartilage, tendons and ligaments, muscle, adipose tissue, and skin (dermis). The proper function of these tissues is predominantly dependent upon the proper proliferation, differentiation, and function of the mesenchymal cells which make up the tissue. Recent advancements in primary human mesenchymal cell culture have led to remarkable progress in the study of these tissues. Landmark experiments have now demonstrated a stem cell basis for many of these tissues, and, furthermore, significant plasticity and inter-conversion of stem cells between these tissues, resulting in a great deal of contemporary excitement and controversy. Newly-developed mesenchymal cell culture techniques have even lead to novel clinical practices for the treatment of disease. The specialized state-of-the-art culture and assay techniques required to grow, differentiate, and assay human mesenchymal cell types for research and clinical applications are described in detail in this volume. In addition, the procurement and processing of human cells from these diverse tissue types are addressed to guide scientists in this area of research.

Manfred R. Koller

Foreword to the Series

This series of volumes is in celebration of Human Cell Culture. Our ability to grow nearly every type of normal and diseased human cell in vitro and reconstruct tissues in 3 dimensions has provided the model systems on which much of our understanding of human cell biology and pathology is based. In the future, human cell cultures will provide the tools for tissue engineering, gene therapy and the understanding of protein function. The chapters in these volumes are written by leading experts in each field to provide a resource for everyone who works with human cells in the laboratory.

John Masters and Bernhard Palsson

Chapter 1

Articular Cartilage

Ross Tubo and Liesbeth Brown
Genzyme Corporation, Tissue Repair Division, One Mountain Road, Framingham, MA 01701. Tel: 001-508-872-8400; Fax: 001-508-872-9080; E-mail: rtubo@genzyme.com

1. INTRODUCTION

Hyaline articular cartilage is composed of chondrocytes encased within the complex extracellular matrix that they produce. The unique biochemical composition of this extracellular matrix is not only responsible for the smooth, gliding, nearly frictionless motion of the articulating surfaces of the knee joint, but also its resistance to compression, withstanding forces of up to five to eight times an individual's body weight. When this highly specialized tissue is damaged by excessive wear or traumatic injury, the tissue cannot elicit an appropriate repair response and the damage may progress to osteoarthritis [15].

The inability of cartilage to repair itself has led to the development of several surgical strategies, the goals of which are to restore patient function by alleviation of the clinical symptoms of pain, locking and swelling associated with articular cartilage damage. The most widely used approaches require physical penetration of exposed subchondral bone by drilling, abrasion, or microfracture [3, 10, 16, 18]. These techniques ultimately result in the formation of cellular repair tissue, known as fibrocartilage, within the cartilage defect site. While fibrocartilage formation provides for symptomatic relief in the near term (1-4 years), the durability of the repair is inferior to that of normal hyaline articular cartilage, and is a likely consequence of the distinct biochemistry of each tissue.

Hyaline articular cartilage is composed of predominately type II collagen and sulfated proteoglycans in a hydrated gel matrix structure [9, 20]. Type II collagen is coiled around a central core of type XI collagen that serves as the primary building block for the organization of type II collagen fibrils within the articular cartilage. Type XI collagen provides the covalent linkage between type II collagen fibrils and appears to cross-link the collagen fiber

M.R. Koller, B.O. Palsson and J.R.W. Masters (eds.), Human Cell Culture Vol V, 1–16.

structure and cartilage specific proteoglycans. The cross-linked collagen fibril and proteoglycan network conveys the properties of tensile and compressive strength respectively to articular cartilage, [see Buckwalter and Mow (13) for review]. In contrast, fibrocartilage is composed predominately of type I collagen and non-cartilage specific proteoglycans and lacks the specific tensile and compressive strength required for long-term function in the joint.

Since hyaline articular chondrocytes are responsible for the production of hyaline articular cartilage *in vivo*, it follows that articular chondrocytes may be a useful cell type for articular cartilage repair. Cultured hyaline articular chondrocytes, isolated from an individual's own cartilage and propagated *ex vivo*, have been used to repair clinically significant defects in articular cartilage (Figure 1). Data from these studies indicate that the repair tissue was judged to be "hyaline-like" [22], characterized by randomly organized chondrocytes in lacunae embedded within an extracellular matrix composed of type II collagen and sulfated proteoglycans [7, 20]. Autologous chondrocyte implantation (ACI) was approved for treatment of articular cartilage defects by the Center for Biologics Evaluation and Research of the Food and Drug Administration in August, 1997.

Articular chondrocytes harvested from small biopsies of articular cartilage removed during arthroscopic examination of the cartilage defect express articular cartilage-specific extracellular matrix components. Once the cells are freed from the tissue by enzymatic digestion and placed into monolayer tissue culture for proliferative expansion, they become phenotypically unstable, adopting a fibroblastic morphology and then ceasing production of type II collagen and cartilage specific proteoglycan. These "de-differentiated" cells proliferate rapidly and produce type I collagen, the collagen characteristic of fibrous tissue. However, when placed in an appropriate environment, such as suspension culture *in vitro* [2, 4, 27], or in the environment of a cartilage defect *in vivo* [25], the cells re-differentiate and express articular cartilage-specific matrix molecules again. The reversibility of de-differentiation is key to the successful repair of articular cartilage using cultured autologous chondrocytes.

The development of reproducible methods for isolating and culturing autologous human articular chondrocytes while maintaining their ability to re-express the hyaline articular cartilage phenotype are central to the successful clinical application of this cell-based technology. Mammalian cell culture processes typically employ a commercially available cell culture medium base supplemented with animal-derived serum for optimal growth [12]. Human articular chondrocytes are typically cultured in Dulbecco's Modified Eagle's Medium (DMEM) supplemented with 10% (v/v) bovine serum [2, 6]. The observation that different samples of serum vary

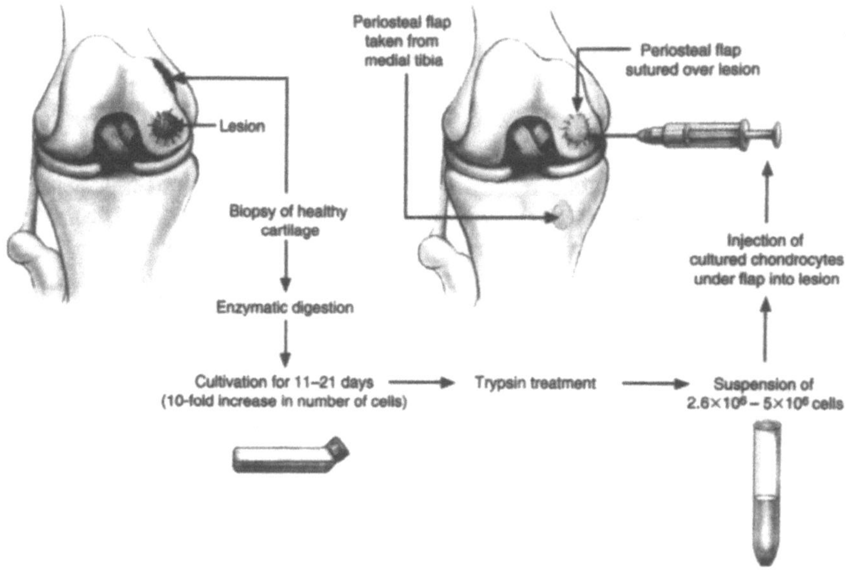

Figure 1. Schematic diagram for articular chondrocyte harvest and implantation, as performed in the treatment of deep cartilage defects in the knee with autologous chondrocyte implantation. From Brittberg et al., 1994.

significantly in their ability to promote proliferation and/or differentiation highlights the potential problems associated with its use in large-scale cell culture processes [28].

In this Chapter, we discuss the development of reproducible methods for the large-scale manufacture of human articular chondrocytes, and demonstrate the utility of *in vitro* models of chondrocyte re-differentiation in the development of these methods.

2. TISSUE PROCUREMENT AND PROCESSING

The procurement of normal healthy human hyaline articular cartilage for research purposes represents a significant challenge to investigators, due in part to the fact that hyaline articular cartilage cannot repair itself once damaged. The removal of such tissue is likely to inflict irreparable damage to the articular cartilage. Therefore, the option of obtaining articular chondrocytes from living donor biopsies harvested from healthy articular cartilage tissue is not feasible.

Articular cartilage for research purposes is typically obtained from two primary sources: tissue banks, which provide tissue from fresh cadavers; and orthopedic surgeons, who remove dislodged pieces of cartilage during arthroscopic surgery. Each of these sources requires a significant amount of preparation prior to receiving tissue, including: assurance of donor confidentiality; favorable review by institutional review boards (IRBs); and comprehensive training of individuals who will harvest and ship the tissue. The inability to regulate the timing of tissue donation and the variable disease state of procured cartilage tissue, particularly for cadaver tissue, are significant hurdles to the regular scheduling of research studies.

Human articular cartilage for our studies was supplied by the National Disease Research Interchange tissue bank (Philadelphia, PA) from multiple donor cadavers (ranging from 5 to 63 years old), isolated within 48 hours post mortem. Articular cartilage was surgically removed by excision from the femoral condyles and harvested cartilage was transported in physiological saline or commercially available serum-free medium within 4 days at 2-8°C. Chondrocytes within articular cartilage maintained good cell viability, up to five days, when stored at 4°C. Alternatively, cartilage fragments were obtained by orthopedic surgeons removing small pieces of tissue during arthroscopic examination of injured knees.

Human articular cartilage samples were trimmed of extraneous material, minced, and subjected to enzymatic digestion for chondrocyte isolation. While a variety of enzymes have been used for the purpose of chondrocyte isolation, including collagenase, hyaluronidase, pronase, deoxyribonuclease, and trypsin [2, 6, 7, 27], collagenase has become the primary enzyme for cell isolation.

3. ASSAY TECHNIQUES

Given that the mechanical properties and durability of hyaline articular cartilage are a function of its extracellular matrix composition and organization, the "tissue engineered" cartilage produced by implanted cells should exhibit the composition and structure of normal articular cartilage. Retention of cultured articular chondrocytes' ability to re-differentiate and produce an extracellular matrix that closely approximates the character of native articular cartilage following proliferative tissue culture can be demonstrated using agarose or alginate suspension cultures [2, 4-6]. These semi-solid media prevent articular chondrocytes from attaching to a substratum, thus stimulating the re-expression of the differentiated cartilage phenotype. When suspended in agarose gels or alginate beads for three to four weeks, de-differentiated chondrocytes form colonies of cells with halos of newly synthesized cartilaginous matrix. Immunohistochemical and molecular analysis of chondrocyte re-differentiation has shown that expanded articular chondrocytes re-express the biochemical and molecular markers characteristic of hyaline articular cartilage [5].

3.1 Chondrocyte re-differentiation in semi-solid media

The ability of cultured chondrocytes to re-differentiate was assessed by three-dimensional suspension culture assay, either in agarose [4], or in alginate [6].

3.2 Agarose differentiation assay

Articular chondrocytes were harvested by trypsinization following 12-14 days of monolayer culture. Trypsin was neutralized with either 10% (v/v) fetal bovine serum (FBS) for serum-supplemented cultures, or 1% (w/v) human alpha-1-anti-trypsin for serum-free cultures. Cells released from their substratum were then suspended at 2.5 x 10^5 cells per ml in DMEM supplemented with FBS. The medium containing cells was then mixed 1:1 with 4% low-melt agarose at 42°C. The low-melt agarose/cell suspension was overlaid onto a layer of high-melt agarose (2ml, previously solidified) in 60mm tissue culture dishes. After solidification of the low-melt agarose containing cells, the cultures were overlaid with 5ml DMEM/10% FBS. Culture medium was replenished following 2-3 hours of equilibration and every 2-3 days thereafter until fixation. After 3-4 weeks, the cultures were fixed in 10% formalin and stained with safranin-O. Safranin-O positive colonies of ≥2 μm diameter were counted using a microscope. For each corresponding monolayer condition, a total of 10 grids of $4mm^2$ each were counted.

3.3 Alginate differentiation assay

Monolayer cultured, de-differentiated chondrocytes were resuspended in alginate solution containing 1.2% (w/v) potassium alginate (KELCO, Rahway, NJ) in 0.15M sodium chloride with 25mM HEPES buffer at pH 7.0, at 10^6 cells/ml. Cells in unpolymerized alginate solution were loaded into a 10cc syringe with a 22 gauge needle, and added by dropwise expulsion into a 0.1M calcium chloride solution. The alginate was allowed to polymerize for 10 minutes in the calcium chloride solution, three-dimensionally entrapping the cells in droplets of polymerized alginate. Cell seeded beads were washed twice in 0.15M sodium chloride with 25mM HEPES buffer at pH 7.0, to remove excess calcium, and then transferred to fresh DMEM supplemented with 10% FBS. The medium was changed every 2-3 days.

3.4 Analysis of gene expression

Re-differentiation of chondrocytes cultured under a variety of conditions was analyzed by RNase protection assay [5]. Five ml aliquots of alginate beads, corresponding to approximately 5 x 10^6 cells, were collected for RNA and immunohistochemical analysis.

3.5 RNA Analysis

Total cellular RNA was prepared from frozen pellets of approximately 5 x 10^6 cells using the RNeasy™ kit (Qiagen, Chatsworth, CA) according to the manufacturer's instructions. Typically, a yield of 5 to 15μg of total RNA was obtained. The purified RNA preparation was either frozen in water or kept as an ethanol suspension at –80°C. All reagents were free of RNase. *In vitro* transcription reactions were performed in order to obtain radiolabeled antisense RNA probes specific for human type I, type II and type X collagens, as well as aggrecan and versican [11, 19, 21, 23, 29]. These probes were used for RNase protection assay.

3.6 RNase protection assay (RPA)

RNase protection assays were performed using the Ambion Hybspeed™ RPA kit (Austin TX). Briefly, one microgram of total RNA was co-precipitated with an aliquot of the purified probes (25,000 cpm per probe) by the addition of 0.1 volume of 5M NH_4O-acetate and 2.5 volume of ethanol. After a 10 minute centrifugation, the pellet was washed with 70% ethanol, dried and solubilized in 10μl of hybridization buffer. Radiolabeled antisense probes and sample RNA were allowed to hybridize for 10 minutes at 68°C, and 100μl of a 1:100 dilution of RnaseA/RnaseT1 mix in digestion buffer was added. The digestion was allowed to proceed for 30 minutes at 37°C before being terminated with the addition of inactivation/precipitation buffer. The samples were centrifuged for 10 minutes and the pellets redissolved in 10μl of sample buffer. The protected fragments were resolved by electrophoresis on an 8M urea containing 4% polyacrylamide gel. The results were visualized by autoradiography.

3.7 Immunohistochemical Analysis

Primary antibodies, T40202R (Biodesign, Kennebunkport, ME) which recognizes type I collagen, T40311R (Biodesign) for type II collagen, MO-225 (Seikagaku, Rockville, MD) for chondroitin SO_4, or normal rabbit IgG, were diluted 1:50 in phosphate buffered saline (PBS) and used to cover cells fixed on microscope slides. The reaction was carried out in a humid atmosphere at 37°C for one hour. The slides were then washed three times in PBS followed by incubation with a 1:200 dilution of rhodamine-conjugated goat anti-rabbit IgG in PBS as a secondary antibody, under the same conditions described for the primary antibodies. Hoechst dye at 1 μg/ml was included in some experiments with the secondary antibodies for nuclear staining. The slides were washed three times in PBS, mounted, and examined with a fluorescence microscope.

4. CULTURE TECHNIQUES

4.1 Anchorage Dependent Expansion in Monolayer

The primary objective for the propagation of any normal human cell type is to develop standardized methods of tissue culture which consistently produce high yields of functional cells. This objective requires special emphasis when considering processes for the large-scale manufacture of autologous cells for clinical use, as required for autologous chondrocyte implantation [7]. The impact of changing the methods of cell culture for articular chondrocytes on their ability to function has been tested using the suspension culture assays described in the previous section. The strategy for optimizing the culture system and the results of these studies are discussed in this section.

4.2 Autologous human serum (AHS) versus fetal bovine serum (FBS)

Autologous human articular chondrocytes for clinical use were first propagated in culture medium composed of Ham's F-12 medium supplemented with 15% (v/v) autologous serum [7]. Cells from each individual patient were cultured in medium supplemented with their own serum. The use of multiple lots of individual patient serum for culture of articular chondrocytes from each patient creates a potential for variability in the production of high quality cells, given the variable levels of growth and differentiation observed for cells in different lots of serum [28].

The colony-forming efficiency of ten different strains of human articular chondrocytes cultured in a single lot of FBS was compared to the colony-forming efficiency of the same strains cultured in donor-matched autologous human serum (AHS). Human articular chondrocytes were isolated and cultured by two distinct methods [5, 7]. Biopsies of human articular cartilage were harvested from a minor load-bearing surface of the articular cartilage of the affected knee, and then trimmed, bisected, and transported to the laboratory for cell isolation. Isolated chondrocytes were propagated in monolayer tissue culture in vessels containing either: Ham's F-12 (F-12; Gibco, Grand Island, NY) supplemented with 15% AHS (Method 1; Brittberg et al., 1994); or Dulbecco's Modified Eagle's Medium (DMEM; Gibco) supplemented with 10% FBS (Hyclone, Logan, UT) (Method 2; Binette et al., 1998). All cultures were incubated at 37°C in a humidified 5% CO_2 environment, with medium changes every 2-3 days. Cells were passaged at 80-90% confluence using 0.05% trypsin-EDTA and were diluted ten-fold for subculture. Trypsin was neutralized by addition of the appropriate serum (10% v/v).

Following approximately three weeks of monolayer culture, chondrocytes cultured by both methods were tested for their ability to re-differentiate using the agarose suspension culture assay, as described previously [4]. The colony-forming efficiencies of chondrocytes cultured in medium supplemented with the single lot of FBS were at least as good as, if not better than the colony-forming efficiencies observed for the same patient strains grown in their own AHS (Figure 2).

4.3 Impact of Donor Age on Chondrocyte Re-differentiation

The culture system developed for large-scale manufacture of autologous chondrocytes for clinical use must be robust in its ability to support proliferation, and ultimately re-differentiation of cells derived from patients of varied age. Therefore, the re-differentiation of chondrocytes derived from the articular cartilage of individuals ranging in age from 5-65 years was examined by agarose assay. Chondrocytes from each individual were cultured in monolayer for at least three weeks in DMEM supplemented with 10% (v/v) FBS, harvested, and then seeded into agarose suspension culture for assessment of colony forming ability (Figure 3).

4.4 Development of a defined medium for articular chondrocyte culture

Human articular chondrocytes were initially seeded at 3-4000 cells per cm^2 and allowed to attach to tissue culture plastic for one day in DMEM supplemented with 10% FBS. Following overnight incubation, culture medium was removed and replaced with DMEM/FBS as control, or the defined medium, as described below.

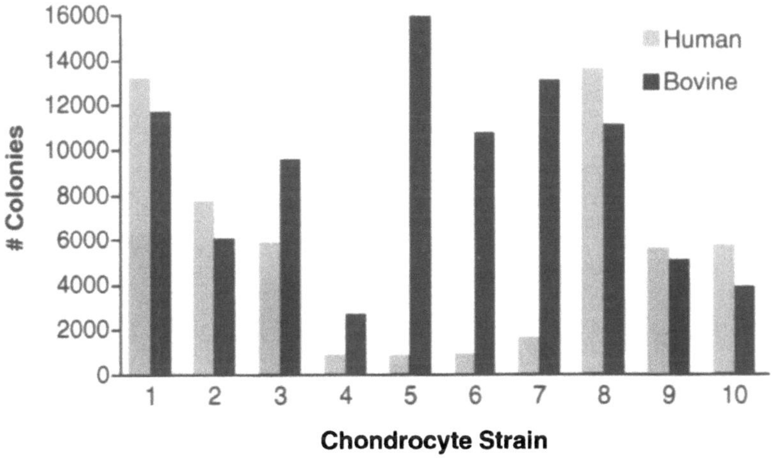

Figure 2. Comparison of chondrocyte culture in autologous human serum versus fetal bovine serum. Human articular chondrocytes were cultured in medium supplemented with fetal bovine serum or donor-matched autologous human serum. Re-differentiation of chondrocytes was assessed by counting colonies in agarose suspension culture as described in Section 3.

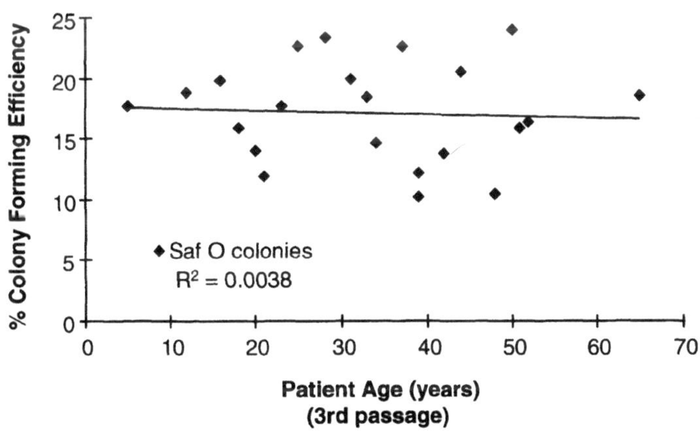

Figure 3. Effect of patient age on the ability of chondrocytes to form colonies in suspension culture following proliferative expansion in monolayer. The colony-forming ability of human articular chondrocytes derived from the articular cartilage of 22 donors ranging in age from 5 to 65 years was assessed following monolayer culture into the third passage (approximately 7-10 population doublings).

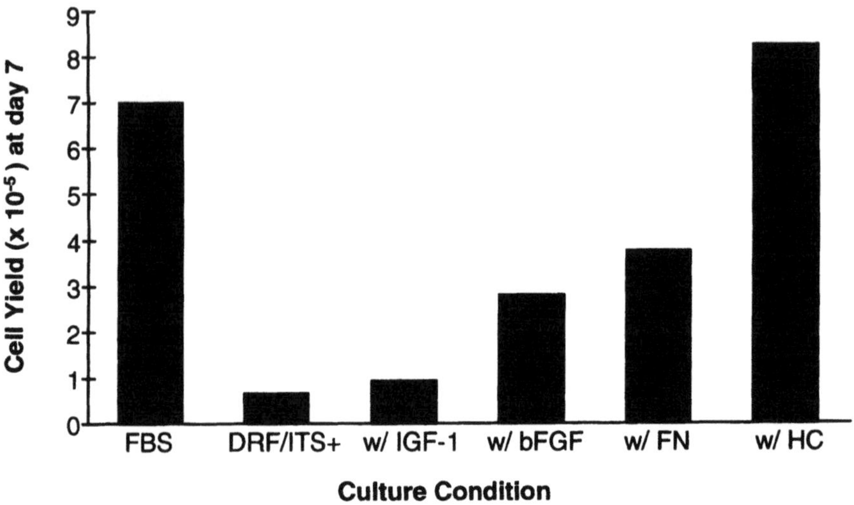

Figure 4. Figure 4: Cell yield following culture of human articular chondrocytes in different formulations of defined medium. Cell yields were compared following culture of normal human articular chondrocytes in (A) serum containing DMEM/FBS; or serum-free DRF medium supplemented with (B) insulin, transferrin, and selenium (ITS+); (C) ITS+ and insulin-like growth factor-1 (IGF-1); (D) ITS+, IGF-1, and basic fibroblast growth factor (bFGF); (E); ITS+, IGF, bFGF, and fibronectin (FN); or (F) ITS+, IGF, bFGF, FN, and hydrocortisone (HC).

Defined medium was composed of a 1:1:1 mixture of three basal media: DMEM, RPMI-1640, and Ham's F12 + HEPES (Gibco), referred to hereinafter as DRF. DRF was then supplemented with (a) ITS+ (insulin at 10 µg/ml, transferrin at 5.5µg/ml, and selenium at 7ng/ml), linoleic acid (5µg/ml), ethanolamine (2µg/ml), and bovine serum albumin (1mg/ml); (b) ITS+, with and without insulin-like growth factor-I (IGF-I); (c) ITS+, IGF-I, with and without basic fibroblast growth factor (bFGF); (d) ITS+, IGF-I, bFGF, with and without fibronectin; or (e) ITS+, IGF-I, bFGF, fibronectin, with and without hydrocortisone. The culture medium in serum-containing cultures was completely replenished every 2-3 days, whereas cultures containing defined medium were not replenished. The cell yields obtained using complete defined medium were equivalent to those obtained in control cultures (Figure 4).

Articular chondrocytes cultured in DMEM supplemented with 10% FBS grew to confluence within 7-9 days (Figure 5A). The morphology of cells cultured in DRF with ITS+, and with ITS+ supplemented with IGF-I, was characterized by large cells having thin processes (Figures 5B and 5C). The addition of bFGF and fibronectin enhanced chondrocyte morphology, with smaller, more numerous, refractile cells (Figures 5D and 5E). Finally, the addition of hydrocortisone to DRF supplemented with ITS+, IGF-I, bFGF, and fibronectin resulted in the achievement of culture confluence and a morphology which was indistinguishable from control, DMEM with 10% FBS (Figure 5F).

4.5 Re-differentiation of articular chondrocytes cultured in defined medium

Human articular chondrocytes were harvested from monolayer culture by trypsinization using 0.05% trypsin and 0.5mM EDTA or, for defined medium cultures, trypsin was neutralized using recombinant 1.0% human alpha-1-anti-trypsin. Cells were seeded directly into either defined medium or DMEM supplemented with 10% (v/v) FBS as control. The ability of chondrocytes to re-differentiate in suspension culture, following proliferative expansion in complete defined medium, was equivalent to that from serum-containing control cultures (Table 1).

Table 1. Re-differentiation potential (%) of human articular chondrocytes following proliferative expansion in defined medium.

Donor Age (years)	Medium for Proliferative Expansion	
	DMEM w/10% FBS	Defined Medium
24	28.5	31.6
31	20.3	24.9
35	22.5	20.5

5. UTILITY OF THE SYSTEM

The clinical application of autologous chondrocyte implantation to the treatment of articular cartilage defects has stimulated the development of highly reproducible, standardized methods for human articular chondrocyte isolation and culture. The autologous nature of the chondrocyte implantation procedure requires that the cell culture medium must promote rapid proliferation and maintenance of re-differentiation potential of adult human articular chondrocytes in patients of varied health and age.

The articular chondrocyte culture system developed by Brittberg et al. (1994) used serum derived from each individual patient as a culture supplement. The inconsistent maintenance of articular chondrocyte re-differentiation potential, measured by suspension culture analysis, in these matched donor "lots" of AHS is the likely result of varying levels of growth and differentiation factors present in different lots of serum [12, 28]. This variability was reduced when the same human articular chondrocyte strains were cultured in a single lot of FBS (Figure 2).

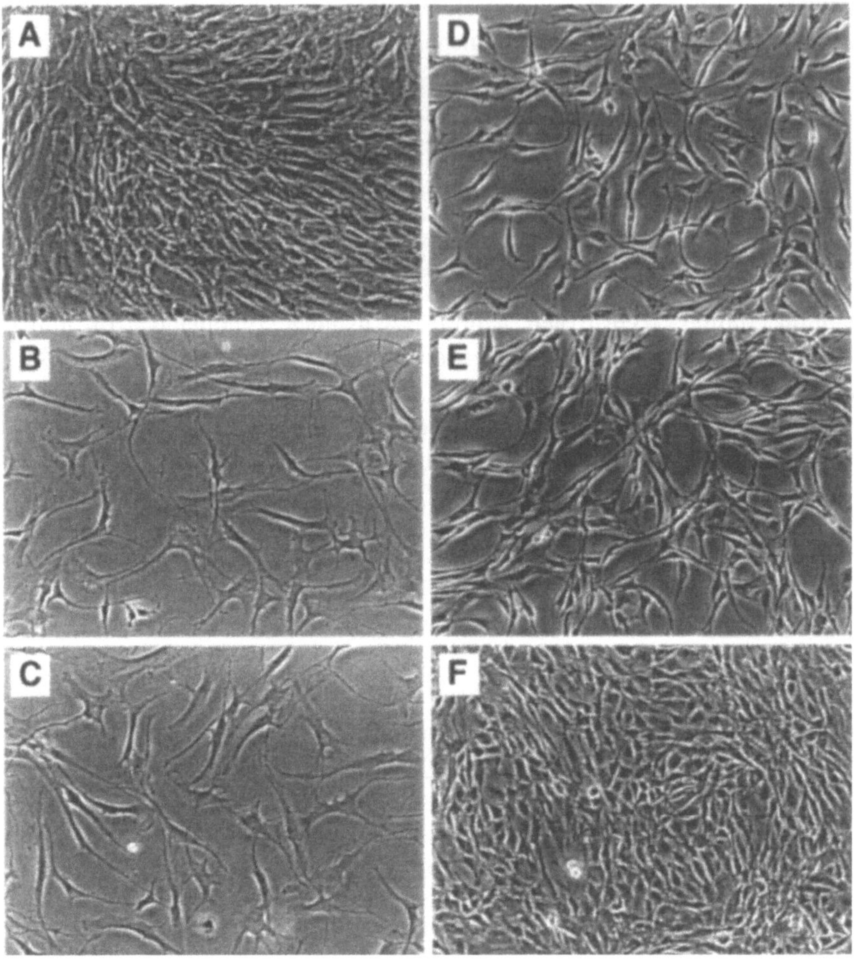

Figure 5. Phase contrast photomicrographs of normal human articular chondrocytes cultured in (A) serum-containing DMEM/FBS; or serum-free DRF medium supplemented with (B) insulin, transferrin, and selenium (ITS+); (C) ITS+ and insulin-like growth factor-1 (IGF-1); (D) ITS+, IGF-1, and basic fibroblast growth factor (bFGF); (E); ITS+, IGF, bFGF, and fibronectin (FN); or (F) ITS+, IGF, bFGF, FN, and hydrocortisone (HC).

Although serum has been widely used for mammalian cell culture, there are several potential problems associated with its use: 1) serum contains many unidentified or non-quantified components and therefore is not "defined"; 2) the composition of serum varies from lot to lot, making standardization difficult for experimentation or other uses of cell culture; 3) because many of these components affect cell attachment, proliferation, and differentiation, controlling these parameters, or studying the specific requirements of cells with respect to these parameters, is precluded by the use of serum; 4) some components of serum are inhibitory to the proliferation of specific cell types, and to some degree may counteract its proliferative effect resulting in sub-optimal growth; and 5) serum may contain adventitious agents that could affect the outcome of experiments or provide a potential health hazard if the cultured cells are intended for implantation in humans [12].

The potential problems associated with culturing chondrocytes in serum-containing medium can be addressed by using serum that is extensively screened for safety and efficacy, as is currently done, or using a defined serum-free medium. The development of defined medium must be tailored to the cell type of interest since requirements for specific nutrients, growth and attachment factors, hormones and other components vary from one cell type to another.

Chondrocytes produce and secrete factors that promote their own attachment and proliferation [24]. Examples include bFGF [14], insulin-like growth factors [13], transforming growth factor-β [26], vitronectin, and possibly some unidentified factors that promote their attachment and proliferation. Maximal cell proliferation and maintenance of redifferentiation capacity were achieved when serum-free, chemically defined medium (DRF) was supplemented with IGF-I, bFGF, hydrocortisone, and fibronectin (Figures 4 and 5).

Previous attempts to culture articular chondrocytes in defined medium have been partially successful [1, 17]. However, exposure to undefined bovine serum was necessary in these systems for 24-48 hours for chondrocyte attachment. While the addition of purified human fibronectin to the defined medium facilitated chondrocyte attachment, the use of undefined serum was slightly superior. Finally, there was no difference in the ability of human articular chondrocytes expanded in defined medium to re-differentiate, as compared to the same human articular chondrocyte strains expanded in FBS-containing medium (Table 1).

6. CONCLUDING REMARKS

6.1 The Future Leads to Tissue Engineering

Tissue engineering has been proposed as a solution for complex clinical problems ranging from structural repair of localized tissue damage to restoration of systemic physiology. The successful treatment of such varied clinical problems with tissue engineered products depends upon the ability of tissue engineered products (i.e. cells, matrices, and/or growth factors) to function in a clinically relevant manner.

The current autologous chondrocyte implantation procedure, with cells delivered in suspension, has a clinical success rate of 80-85% for single contained cartilage defects on the femoral condyle in an otherwise healthy knee [8, 20, 22]. However, patient recovery and return to full physical activity may take up to twelve months, due to the invasiveness of the surgical procedure and the time required for articular chondrocytes to re-differentiate and produce functional hyaline articular cartilage matrix [5]. Therefore, the combination of articular chondrocytes with a biocompatible extracellular matrix delivery vehicle, delivered in a non-invasive manner and supplemented with cell growth and differentiation enhancing agents, represents an attractive strategy for articular cartilage repair. The maintenance of the articular chondrocytes' capacity to re-differentiate in these tissue engineered constructs will help to guide the future of hyaline articular cartilage repair in the knee.

REFERENCES

1. Adolphe, MBF, Ronot X, Corvol M, and Forest N (1984) Cell multiplication and type II collagen production by rabbit articular chondrocytes cultivated in a defined medium. *Exp. Cell Res.* 155:527-536.
2. Aulthouse, AL, Beck M, Griffey E, Sanford J, Arden K, Machado MA, and Horton WA (1989) Expression of the human chondrocyte phenotype in vitro. *In Vitro Cellular & Developmental Biology*. 25:659-668.
3. Baumgaertner, MR, Cannon WD, Jr., Vittori JM, Schmidt ES, and Maurer RC (1990) Arthroscopic debridement of the arthritic knee. *Clinical Orthopaedics & Related Research*:197-202.
4. Benya, PD, and Shaeffer JD (1982) Dedifferentiated chondrocytes reexpress the differentiated collagen phenotype when cultured in agarose gels. *Cell*. 30:215-224.
5. Binette, F, McQuaid DP, Haudenschild DR, Yaeger PC, McPherson JM, and Tubo R (1998) Expression of a stable articular cartilage phenotype without evidence of hypertrophy by adult human articular chondrocytes in vitro. *Journal of Orthopaedic Research*. 16:207-16.
6. Bonaventure, J, Kadhom N, Cohen-Solal L, Ng KH, Bourguignon J, Lasselin C, and Freisinger P (1994) Reexpression of cartilage-specific genes by dedifferentiated human articular chondrocytes cultured in alginate beads. *Experimental Cell Research*. 212:97-104.
7. Brittberg, M, Lindahl A, Nilsson A, Ohlsson C, Isaksson O, and Peterson L (1994) Treatment of deep cartilage defects in the knee with autologous chondrocyte transplantation. *New England Journal of Medicine*. 331:889-95.
8. Browne, JE, Fu FH, Mandelbaum BR, Micheli LJ, and Moseley B (1998) Autogenous chondrocyte implantation for treatment of articular cartilage knee defects: 2-4 year multi-center experience. *In* Transactions of the Eighth ESSKA Congress, Nice, France.
9. Buckwalter, JA, Rosenberg LC, and Hunziker EB. 1990. Articular Cartilage: Composition, Structure, Response to Injury, and Methods of Facilitating Repair. *In* Articular Cartilage and Knee Joint Function: Basic Science and Arthroscopy. JW Ewing, editor. Raven Press Ltd., New York. 19-56.

10. Chang, RW, Falconer J, Stulberg SD, Arnold WJ, Manheim LM, and Dyer AR (1993) A randomized, controlled trial of arthroscopic surgery versus closed-needle joint lavage for patients with osteoarthritis of the knee. *Arthritis & Rheumatism*. 36:289-96.
11. Doege, KL, Sasaki M, Kimura T, and Yamada Y (1991) Complete coding sequence and deduced primary structure of the human cartilage large aggregating proteoglycan, Aggrecan. *Journal Biology Chemestry*. 266:894-902.
12. Freshney, I. (1994). Serum-free media. *In* Culture of Animals Cells. John Wiley & Sons, New York. 91-9.
13. Froger-Gaillard, B, Hossenlopp P, Adolphe M, and Binoux M (1989) Production of insulin-like growth factors and their binding proteins by rabbit articular chondrocytes: relationships with cell multiplication. *Endocrinology*. 124:2365-72.
14. Hill, DJ, Logan A, Ong M, De Sousa D, and Gonzalez AM (1992) Basic fibroblast growth factor is synthesized and released by isolated ovine fetal growth plate chondrocytes: potential role as an autocrine mitogen. *Growth Factors*. 6:277-94.
15. Hunter, W (1743) On the structure and diseases of articulating cartilage. *Philos Trans R Soc Lond*. 42B:514-521.
16. Jackson, RW. 1991. Arthroscopic treatment of degenerative arthritis. *In* Operative Arthroscopy. J McGinty, editor. Raven Press, New York, NY. 319-323.
17. Jennings, SD, and Ham RG (1983) Clonal Growth of primary cultures of human hyaline chondrocytes in a defined medium. *Cell Biology International Reports*. 7:149-159.
18. Johnson, LL. 1991. Arthroscopic abrasion arthroplasty. *In* Operative Arthroscopy. J McGinty, editor. Raven Press, New York, NY. 341-360.
19. Kuivaniemi, H, Tromp G, Chu ML, and Prockop DJ (1988) Structure of a full-length cDNA clone for the prepro alpha2 (I) chain of human type I procollagen. Comparison with the chicken gene confirms usual patterns of gene conservation. *Biochem. J*. 252:633-640.
20. Minas, T, and Nehrer S (1997) Current concepts in the treatment of articular cartilage defects. *Orthopedics*. 20:525-538.
21. Muragaki, Y, Kimura T, Ninomiya Y, and Olsen B (1990) The complete primary structure of two distinct forms of human alpha 1 (IX) collagen chains. *Eur J Biochem*. 192:703-708.
22. Peterson, L (1998) Autologous chondrocyte transplantation 2010 year follow-up in 219 patients. *In* American Academy of Orthopaedic Surgery 65th Annual Meeting, New Orleans, LA.
23. Reichenberger, E, Beier F, Valle PL, Olsen B, Mark KVd, and Bertling W (1992) Genomic organization and full length cDNA sequence of human collagen X. *FEBS Lett*. 311:305-310.
24. Shen, V, Rifas L, Kohler G, and Peck WA (1985) Fetal rat chondrocytes sequentially elaborate separate growth- and differentiation-promoting peptides during their development in vitro. *Endocrinology*. 116:920-5.
25. Shortkroff, S, Barone L, Hsu HP, Wrenn C, Gagne T, Chi T, Breinan H, Minas T, Sledge CB, Tubo R, and Spector M (1996) Healing of chondral and osteochondral defects in a canine model: the role of cultured chondrocytes in regeneration of articular cartilage. *Biomaterials*. 17:147-154.
26. Villiger, PM, Kusari AB, ten Dijke P, and Lotz M (1993) IL-1 beta and IL-6 selectively induce transforming growth factor-beta isoforms in human articular chondrocytes. *Journal of Immunology*. 151:3337-44.
27. Watt, FM, and Dudhia J (1988) Prolonged expression of differentiated phenotype by chondrocytes cultured at low density on a composite substrate of collagen and agarose that restricts cell spreading. *Differentiation*. 38:140-147.

28. Yaeger, PC, Masi TL, de Ortiz JL, Binette F, Tubo R, and McPherson JM (1997) Synergistic action of transforming growth factor-beta and insulin-like growth factor-I induces expression of type II collagen and aggrecan genes in adult human articular chondrocytes. *Experimental Cell Research.* 237:318-25.
29. Zimmermann, DR, and Ruoslahti E (1989) Multiple domains of the large fibroblast proteoglycan, versican. *EMBO J.* 8:2975-2981.

Chapter 2

Tendon and Ligaments

Louis C Almekinders and Albert J Banes
Department of Orthopaedic Surgery, University of North Carolina CB# 7055, Chapel Hill, NC 27599. Tel: 001-919-966-9079; Fax: 001-919-966-6730; E-mail: almekind@med.unc.edu

1. INTRODUCTION

The term "musculoskeletal system" suggests that muscles and bones are the key components of this system. However, without ligaments and tendons, the supportive and locomotor functions of this system would be impossible to use.

Tendons serve as connections between muscle and bone, thereby allowing the muscles to exert their contractile action. Most muscles in the human body have a tendon between the muscle and its distal bony attachment site. In some instances, the proximal part of the muscle appears to connect directly to the bone. The shape and size of the tendon can vary greatly. Flat and broad muscles tend to have flat and short tendons, such as the deltoid muscle of the shoulder. Long and often slender muscles tend to have long and round tendons such as finger flexor tendons.

Ligaments provide stability for a joint as the muscle-tendon units can move that joint through a certain range of motion. Most ligaments can be found outside the joint cavity as separate structures or as thickened bands in the capsule. Some ligaments have an intra-articular location and are entirely within the joint cavity. The anterior cruciate ligament of the knee is an example of an intra-articular ligament.

For a long time, both tendons and ligaments have been considered rather passive structures with simple anatomic features. During the past two decades, there has been a remarkable increase in research interest in ligaments and tendons largely because of the high prevalence of injuries to these structures. Tendon and ligament injuries comprise the majority of all musculoskeletal injuries seen both in sports and occupational medicine [1].

M.R. Koller, B.O. Palsson and J.R.W. Masters (eds.), Human Cell Culture Vol V, 17–25.

Acute ruptures, often called sprains or strains, as well as chronic conditions such as tendonitis, are common in both young and elderly patients.

Tendons and ligaments are not simple and straightforward structures as previously thought. Under light microscopy, tendons and ligaments appear somewhat similar, with wave-like extracellular matrix comprised mainly of Type I collagen in the load bearing tendon fascicles and Types I and III collagens in the epitenon [2]. Fibrillar collagen molecules are long triple helix proteins that are crosslinked and give the tendons and ligaments their ability to withstand tensile forces. Interspersed between the fibrils are syncytial fibroblasts that produce and maintain this extracellular matrix. More in depth investigations have revealed clear differences, however, between as well as within ligaments and tendons [3]. Collagen fibril diameter measurements have shown predominantly large diameter fibrils in tendons as opposed to smaller fibrils produced by the ligament fibroblasts. Tendon fibroblasts appear to consist of different subpopulations. The outer layer of the tendon or epitenon contains surface fibroblasts that can be isolated separately [4]. The core of the tendon contains internal fibroblasts (IF) that have distinctly different properties when compared to the tendon surface cells (TSC). This is most likely related to their physiologic function, as the TSC appear to be involved with maintaining the gliding surface of the tendon and respond first to traumatic injury by migrating to the wound site, and then dividing.

Within ligaments and tendons themselves, significant differences exist. Some tendons course around bony prominences at rather sharp angles. At those sites, significant compression is exerted on the tendon in addition to the tensile forces. Metaplasia to cartilage-like extracellular matrix with chondrocyte-like cells can be found at those sites [5]. Similar cartilaginous changes have been reported directly adjacent to bone at some bony attachment sites [6]. Cartilaginous transition zones have also been reported at ligament insertion sites [7].

The difference in healing characteristics of intra-articular versus extra-articular ligament has led to research on the differences within ligament fibroblasts. Differences in the response to mechanical stimulation, growth factors and cell adhesion have been found, particularly when comparing medial collateral ligament fibroblasts and anterior cruciate ligament fibroblasts of the knee [8-10].

Both tendon and ligament cells respond to growth factors by cell migration, cell division and matrix expression. Platelet-derived growth factor (PDGF-BB) and epidermal growth factor (EGF) elicit maximal cell migration in tendon cells, whereas insulin-like growth factor (IGF-I) and transforming growth factor-β (TGF-β) stimulate migration in vitro to a lesser extent [11]. Ligament cells have a similar response, except that EGF may be a more potent stimulator for migration. For cell division in tendon cells, PDGF-BB is most potent, followed by EGF, IGF-I and TGF-β. Ligament cells respond similarly. Combinations of PDGF-BB and IGF-I stimulate cell division maximally. For collagen expression in tendon cells, IGF-I and TGF-β are most stimulatory. The same holds for ligament cells.

Research on tendon and ligament fibroblasts is still relatively new. Future research needs to define the tendon and ligament fibroblasts better. Subpopulations are still being found and culture techniques are being developed. This chapter describes current culture techniques and applications.

2. TISSUE PROCUREMENT AND PROCESSING

The procurement of human ligament or tendon tissue is difficult because of the relatively inaccessible location of these structures. A few options exist to procure normal tissue samples from live donors. Knee ligaments are probably the most commonly studied ligaments of the musculoskeletal system. Access to the knee joint in a live donor can be gained through arthroscopic methods. Through small stab incisions, a fiberoptic camera system and instruments can be introduced into the knee under local anesthesia. Both the anterior and posterior cruciate ligaments can be visualized with this method. Small biopsies can be taken without undue damage to the overall integrity of the ligament. The reported use of this technique has focused on biopsies of ruptured anterior cruciate ligaments when the arthroscopy is done as part of the treatment [12]. An arthroscopy done solely for the purpose of obtaining a biopsy is difficult to justify, as this represents a costly procedure with associated surgical risks. A similar procedure could theoretically be done in the shoulder joint with biopsy of the glenohumeral ligaments, although this has not been reported for in vitro studies. Tendons are also relatively inaccessible. Percutaneous biopsy of the Achilles tendon in humans has been reported for histologic evaluation [13]. Presumably, this would also be an option to obtain a sample for culture. Occasionally, normal tendon tissue can be obtained during reconstructive surgery. In particular, tendon grafting and transfer procedures involve the use of normal tendon tissue. Often, excess tendon tissue is discarded, but can be used as an excellent source of tendon fibroblasts. Care should be taken not to use tendon that is located immediately adjacent to bone or near a bony prominence as chondrocyte-like cells may be present in these regions [5, 6].

Specimens for culture of abnormal tissue are generally available because of surgical treatment for these abnormalities. Ruptured tendon and ligaments are often treated with surgical repair or reconstruction. The torn edges of the tendon or ligaments are often removed and discarded, and these can be used as a source of cells. However, tendon ruptures are often due to pre-existing degeneration of the tendon, and the cells that can be cultured from these specimens are not necessarily normal tendon fibroblasts [14]. Rupture of the ligaments is more likely to occur without pre-existing pathology. If specimens are obtained from ruptured ligaments, they can probably be regarded as normal tissue if collected shortly after the rupture and before a significant healing response has occurred.

Fresh cadavers are probably the most common source of tendon and ligament specimens. Technically, any tendon or ligament can be used as a source of cells. Again, it may be important to avoid using tissue immediately adjacent to bone as this often contains a cartilaginous zone [6, 7]. There are no reported studies that indicate the maximum time after death that would still allow harvest of viable fibroblasts. This time can probably be extended if the structure is cooled quickly after death of the donor.

The harvest of specimens should be done under sterile conditions. Initially, the area is cleaned with an antibacterial soap and subsequently wiped with 70% ethanol. Following incision of the skin and dissection of the tendon or ligament, a tissue specimen can be collected and wrapped in moistened sterile gauze in culture medium with antibiotics without serum. If the specimens have to be transported, it is generally done in sterile containers with standard culture medium (e.g. Dulbecco's Modified Eagle's Medium and antibiotics without serum). Prior to explantation, the specimens should be prepared in a sterile hood using sterile techniques.

Ligament specimens may have a synovial covering, particularly if they are intra-articular ligaments. This can be carefully scraped off with a sharp instrument. Large specimens are sectioned transversely in multiple segments to expose more of the core of the ligament. These segments are then cultured using standard tissue culture techniques (see section 4).

Tendon specimens can be treated similarly. However, since there are different subpopulations of tendon fibroblast, it may be important to separate those first [4]. To separate the tendon surface cells (TSC) from internal fibroblasts (IF), the tendon specimens are incubated in 0.5% collagenase in EBSS calcium-containing medium, using culture tubes in a roller apparatus for 10 minutes. The medium is decanted after 10 minutes and replaced by 0.25% trypsin in DMEM-H with HEPES (pH 7.2) and antibiotics for another 10 minutes. Subsequently, the tubes are vortexed for 1 minute and the cells are allowed to settle. Tendon specimens are then removed, leaving mostly TSC that can be plated in serum-containing medium. The removed tendon specimen is used to culture IF. The remaining epitenon is scraped off with a rubber policeman and removed by cutting off each end of the tendon. The specimen is minced with a sterile scalpel into approximately $1mm^3$ pieces and incubated with 0.5% collagenase for 45 minutes at 37°C. The supernatant fluid and cells are sedimented at low speed (900 *g* for 5 minutes). The pellet will contain mostly IF that can be plated in serum containing medium.

3. ASSAY TECHNIQUES

Relatively little is known regarding the specific characteristics of subpopulations of ligament and tendon fibroblasts. Particularly in ligaments, it is not clear whether there are distinct subpopulations. Differences have been found between fibroblasts derived from the medial collateral ligament (MCL), an extra-articular ligament, as compared to those derived from the anterior cruciate ligament (ACL), an intra-articular ligament [15]. Morphologically, ACL fibroblasts appear more hexagonal, whereas MCL cells seem more spindle shaped and elongated. Phalloidin staining revealed a larger number of microfilaments in ACL cells. Proliferation and migration appeared to be significantly lower in ACL cultures. A medium-dependent, differential response to fluid-induced shear stress has also been found [8]. Similarly, different responses to several growth factors have been found in cultured ligament cells [9]. However, these previously mentioned studies have used animal cells, and it is not known whether similar differences exist in human cell types. The only study involving human cells showed a difference in signal pathways upon binding to fibronectin [10]. Using a micropipette-micromanipulation system, the individual cell adhesiveness to a fibronectin-coated glass surface was determined. By using inhibitors of signaling pathways, it was found that MCL fibroblasts play a crucial role in cAMP and Ca^{2+}/phospholipid signaling during integrin-mediated cell adhesion. In ACL fibroblasts, this signal pathway appears to play only a minor role. Whether this truly represents a phenotypic difference, and is therefore usable as an assay technique, is yet to be determined.

Cultured tendon cells, following separation into TSC and IF, can be evaluated [4]. Microscopy has revealed that TSC are relatively large cells with lipid-containing cytoplasmic vesicles. Conversely, IF are smaller, fusiform cells that are more sensitive to trypsin, as compared to TSC. When comparing proliferation rates, TSC have a shorter generation time. Immunohistologic staining for fibronectin, as well as ELISA and mRNA quantitation, have been used to distinguish TSC and IF in culture [16]. Fibronectin levels are markedly increased in TSC culture, as compared to IF culture. TSC and IF have different gene responses to growth factors and other agonists, particularly in response to mechanical loading [17, 18]. Studies on tendon cell subpopulations also involve the use of animal specimens. Although not published, we have been able to confirm similar findings in our laboratory using human specimens. However, because of the difficulties in obtaining fresh, normal human specimens, only a limited number of investigations have been done using human cells.

4. CULTURE TECHNIQUES

Explants, as well as isolated human tendon and ligament fibroblasts, can be readily grown using standard cell culture techniques. Most studies use commercially-available culture dishes or flasks. Viable cells will become adherent to the surface of the dish or flask within 15 minutes to 2 hours. However, over-treatment with collagenase and trypsin upon isolation will compromise attachment, and may even result in cell death. The basic medium used for most studies is Eagle's minimal essential medium or Dulbecco's modified Eagle's medium at neutral pH (7.0 – 7.2). Depending on the purpose of the culture, several additions to the medium may be necessary. For studies extending over several days, standard antimicrobials are generally added to avoid infection of the culture medium. Various combinations are used, often including streptomycin (100 μg/ml), penicillin (100 U/ml) and amphotericin B (0.25 – 2 μg/ml).

Addition of 10% FBS, non-essential amino acids, insulin, transferrin, selenium and L-glutamine (4 mM) are generally used if proliferation of the fibroblasts is desired. Recent studies have shown that proliferation can also be stimulated with different methods. Mechanical stimulation through in vitro stretching of the cells appears to amplify the proliferative response of tendon fibroblasts [19, 20]. The in vitro stretching can be accomplished by growing isolated cells in specially designed culture wells that contain a flexible base. After placing the wells on an airtight rubber gasket, the pressure underneath the well can be reduced in a cyclic manner using a computer-controlled pressure system. This results in a stretching of the flexible substrate and the cells attach to the substrate. Experimental devices to stretch entire tendon segments in vitro have also been used. Those devices are generally custom-built, and have mainly been used for animal tendon specimens [21]. Addition of PDGF-BB or TGF-β has also resulted in dose-dependent proliferative responses in some animal tendon and ligament fibroblasts [22, 23]. No studies on human cells are available in this regard.

Addition of ascorbic acid (0.5 mM) to the medium may be important if the production of collagenous matrix by the cultured cells is an important feature of the study. Ascorbate appears to be an essential factor in this process. Additional stimulation of matrix production in animal ligament fibroblasts has been observed with TGF-β [9].

Several additional substrates are available for use in ligament and tendon fibroblast culture. Collagen-coated culture wells are often used when flexible well bases for mechanical stretching are needed. This allows for a firm adherence of the cells to the base of the well, thereby allowing stretching by as much as 25% compared with the resting length. Similarly, fibronectin-coated surfaces are used to test the adhesiveness of fibroblasts [10].

Fibroblasts usually are cultured under standard conditions. A humidified incubator at 37°C with 5% CO_2 and 95% air is recommended. The culture medium is generally changed every 2 to 4 days to maintain active proliferation. Once confluent cultures are obtained, the cells can be subcultured in multiple wells (usually a 1:2 or 1:3 split) for another passage. In order to split the cells, the culture is first washed several times with PBS at pH 7.2. Subsequently, cultures are treated with trypsin (0.05%) with gentle shaking, collected, washed and diluted, and then re-plated in serum-containing medium in new wells. It is generally assumed that multiple passages will result in some loss of phenotypic features, and although precise data on human fibroblasts are not available, their use is confined to the first 6 passages.

5. UTILITY OF THE SYSTEM

There is a great interest in improving the rate and quality of healing in ligament and tendon injuries. In vivo models are expensive and cumbersome and so the culture of fibroblasts has allowed a relatively simple and economic means of testing the individual effects of growth factors and other substances that theoretically could improve ligament and tendon healing. Once positive effects on proliferation and/or matrix production are found, they can be tested on *in vivo* models. One major limitation of *in vitro* studies is the lack of inflammatory features in cell culture. Most ligament and tendon injuries are thought to have some components of inflammation in vivo. It is difficult to duplicate this environment in a cell culture system. However, recent studies indicate that tendon cells are responsive to inflammatory cytokines.

Another major application of fibroblast cell culture is research into mechanotransduction. Clinically, it has been evident for many years that motion can control many processes in the musculoskeletal system. Both in physiological and pathological states, mechanical stimulation appears to control function and healing. Culture systems have been devised that can exert precise mechanical stimulation of cells or explants. This has allowed study of the beneficial effects of motion, cell adhesion and combination of mechanical stimulation and growth factors.

The most recent research using fibroblast cell culture has been aimed at gene therapy. By introducing viral vectors into cultured fibroblasts, certain genes can be implanted. Expression of genes that control certain cytokines or growth factors can potentially improve the rate and quality of ligament and tendon repair. Reimplantation of these transduced fibroblasts into the injured area could accomplish this goal. Studies in this field are still in their infancy, and the feasibility of such an approach has yet to be proven [24, 25].

6. CONCLUDING REMARKS

Culture of human ligament and tendon fibroblasts is clearly possible with relatively simple culture techniques. Unfortunately, human specimens are not easy to procure, and currently most studies rely on animal data. Much work is yet to be done to fully characterize subpopulations of human fibroblasts. In spite of these limitations, in vitro studies have already yielded interesting findings that could become clinically relevant.

REFERENCES

1. Praemer A, Furner S and Rice DP (1992) Musculoskeletal conditions in the United States. American Academy of Orthopaedic Surgeons, Park Ridge, IL.
2. Tsuzaki M, Yamauchi M and Banes AJ. (1993). Tendon collagens: Extracellular matrix composition in shear stress and tensile components of flexor tendons. Connect Tissue Res 29:141-152.
3. Frank C, Woo S, Andriacchi T et al. (1987) Normal ligament: structure, function, and composition, In Woo S and Buckwalter JA (eds) Injury and repair of musculoskeletal soft tissues. Americian Academy of Orthopaedic Surgeons, Park Ridge, IL.
4. Banes AJ, Donlon K, Link GW et al. (1988) Cell populations of tendon: A simplified method for the isolation of synovial cells and internal fibroblasts. J Orthop Res 6: 83-94.
5. Vogel KG and Koob TJ (1989) Structural specialization in tendons under compression. Int Rev Cytol 115: 267-293.
6. Benjamin M, Evans EJ and Copp L (1986) The histology of tendon attachments to bone in man. J Anat 149: 89-100.
7. Woo SL-Y, Gomez MA, Sites TY at al. (1987) The biomechanical and morphological changes in the medial collateral ligament of the rabbit after immobilization and remobilization. J Bone Joint Surg 69A: 1207-1211.
8. Hung CT, Allen FD, Pollack SR et al (1997) Intracellular calcium response of ACL and MCL ligament fibroblasts to fluid-induced shear stress. Cell Signal 9: 587-594.
9. Marui T, Niyibizi C, Georgescue HI et al (1997) Effects of growth factors on matrix synthesis by ligament fbroblasts. J Orthop Res 15: 18-23
10. Sung K-L P, Whittemore DE, Yang Li et al (1996) Signal pathways and ligament cell adhesiveness. J Orthop Res 14: 729-735.
11. Bynum D, Almekinders L, Benjamin M, Ralphs J, McNeilly C, Yang X, Kenamond C, Weinhold P, Tsuzaki M and Banes AJ. (1997) Wounding in vivo and PDGF-BB in vitro

stimulate tendon surface cell migration and loss of connexin-43 expression. Transactions of the 43rd Annual Meeting of the ORS, Vol. 22 (1), p. 26-5.
12. Spindler KP, Clark SW, Nanney et al (1996) Expression of collagen and matrix metalloproteinases in ruptured human anterior cruciate ligament: an in situ hybridization study. J Orthop Res 14: 857-861.
13. Movin T, Gunter P, Gad A et al (1997) Ultrasonography-guided percutaneous core biopsy in Achilles tendon disorder. Scan J Med Sci Sports 7: 244-248.
14. Kannus P and Jozsa L (1991) Histopathological changes preceding spontaneous rupture of a tendon. J Bone Joint Surg 73A: 1507-1525.
15. Nagineni CN, Amiel D, Green M et al. (1992) Characterization of the intrinsic properties of the anterior cruciate and medial collateral ligament cells: an in vitro cell culture study. J Orthop Res 10: 465-475.
16. Banes AJ, Link GW, Bevin AG et al (1988) Tendon synovial cells secrete fibronectin in vivo and in vitro. J Orthop Res 6: 73-82.
17. Banes AJ, Tsuzaki M, Hu P, Brigman B, Brown T, Almekinders L, Lawrence WT and Fischer T. (1995). Cyclic mechanical load and growth factors stimulate DNA synthesis in avian tendon cells. Special Issue on Cytomechanics in J Biomechanics *28*:1505-1513.
18. Banes AJ, Tsuzaki M, Yamamoto J, Fischer T, Brown T and Miller L. (1995). Mechanoreception at the cellular level: The detection, interpretation and diversity of responses to mechanical signals. Special Issue on Cytomechanics, Biochemistry and Cell Biology *73*:349-365.
19. Almekinders LC, Banes AJ and Ballenger CA (1992) Effects of repetitive motion on human fibroblasts. Med Sci Sports Exerc 25: 603-607.
20. Almekinders LC, Banes AJ and Bracey LW (1995) An in vitro investigation into the effects of repetitive motion and nonsteroidal anti-inflammatory medication on human tendon fibroblasts. Am J Sports Med 23: 119-123.
21. Banes, AJ, Weinhold, P, Yang X, Tsuzaki, M, Bynum D, Bottlang M, and Brown T (2000) Gap junctions regulate responses of tendon cells ex vivo to mechanical loading. Clin Orthop and Rel Res, in press.
22. Banes AJ, Tsuzaki M, Hu P et al (1995) PDGF-BB, IGF-I and mechanical load stimulate DNA synthesis in avian tendon fibroblasts in vitro. J Biomech 28: 1505-1513.
23. Spindler KP, Imro AK Mayes CE et al (1996) Patellar tendon and anterior cruciate ligament have different mitogenic responses to platelet-derived growth factor and transforming growth factor beta. J Orthop Res 14: 542-546.
24. Hildebrand KA, Deie M, Allen CR et al (1999) Early expression of marker genes in the rabbit medial collateral and anterior cruciate ligaments: the use of different viral factors and the effects of injury. J Orthop Res 17: 37-42.
25. Nakamura N, Boorman R, Marchuk L et al (1999) Decorin antisense gene therapy improves early ligament healing – A morphological and high load mechanical assessment. Orthop Res Society Transactions, p. 300.

Chapter 3

Periodontal Ligaments

Thomas Oates Jr and Anh M Hoang
University of Texas Health Science Center at San Antonio, 7703 Floyd Curl Drive, San Antonio, Texas 78284-7894, Tel: 001-210 567-3590. Fax: 001-210-567-6858; E-mail: oates@uthscsa.edu

INTRODUCTION

The periodontal ligament (PDL) is the fibrous connective tissue providing the attachment of the root of a tooth to the surrounding alveolar bone. The PDL is thus the central component of the periodontal attachment apparatus, connecting two mineralized tissues, the root surface cementum and alveolar bone proper. The main function of the PDL is to support the tooth in the bone in response to forces applied to the tooth as may be found during mastication. This supportive function is accomplished through the numerous collagen fiber bundles comprising the bulk of the PDL. These fibers insert into the mineralized tissues and, in response to forces, may stimulate metabolic activity in these mineralized tissues.

The PDL, in its supportive role, appears to be of primary importance in maintaining the homeostatic relationship between the two mineralized tissues.This is most obvious clinically in orthodontic tooth movement. During such controlled tooth movement, on the pressure side of the root in which the PDL is compressed against the osseous tissue, there is bone resorption. In circumstances in which these forces are exerted in excess, there is a tendency for resorption of the mineralized root surface as well as the bone tissue. On the tension side of the root surface in which the PDL is stretched between the two mineralized tissues, there is bone formation. In clinical situations in which forces are applied in more than one direction, the ligament may "adapt" to the forces by widening in multiple directions with compensatory bone resorption circumferentially. In addition to providing functional support, the PDL also has formative, nutritive, and sensory roles. However, at present it is the supportive, and perhaps, the formative roles of

M.R. Koller, B.O. Palsson and J.R.W. Masters (eds.), Human Cell Culture Vol V, 27–41.

this tissue that provide its primary clinical, and therefore research, importance.

Clinically, destructive periodontal disease leads to the degradation of the PDL and adjacent alveolar bone. The progressive loss of these tissues may ultimately lead to loss of the tooth. Therefore, recent therapeutic efforts have focused on the potential to not only inhibit this disease progression, but to actually reverse the degradative process and regenerate the periodontal tissues.

The potential for these tissues to regenerate appears to be selectively associated with the PDL [1]. It is thought that cells from the PDL are critical in their migration and proliferation into the periodontal wound space, and subsequent differentiation into cells capable of reforming all three tissues of the periodontal attachment apparatus. Recent research efforts have focused on understanding and regulating the regenerative potential of cells derived from the PDL.

This leads to the question as to which cells from the PDL are important in this regenerative process. The PDL consists of numerous cell types, including fibroblasts which are the main cells in the tissue, cementoblasts which line the root surface, osteoblasts, epithelial cell rests of Malassez, and undifferentiated mesenchymal stem cells [2]. Although the answer to the question of which cells are important to regeneration is as yet unknown, it has been suggested from kinetic data that perivascular progenitor cells may be critical to the regenerative process [3]. However, it is uncertain whether the perivascular cells are selected through current culturing techniques.

Many of the original studies leading to in vitro culture techniques were related to the development and function of the PDL. The questions ranged from trying to understand the significance of isolated islands of epithelial cells (the epithelial rests of Malassez) within the PDL to potentially seeding the root surfaces of allogenic teeth with recipient cells for transplantation without immune reactions [4, 5]. However, most recent in vitro research has focused on the fibroblastic cells cultured from the PDL as a potential stem cell population leading to regeneration of the periodontal attachment apparatus.

Consistent with fibroblastic cell populations derived from other tissues, these studies have identified heterogeneity in the cell populations cultured [6, 7]. The variety of responses that are related to PDL cells suggest that there may be considerable heterogeneity in cell populations [8-10]. Primary cultures of these fibroblastic cells have been demonstrated to vary in their cellular activities and their responses to specific environmental stimuli. Fibroblastic populations have been shown to vary in their production of total collagen, and in the relative proportions of types I and III collagen produced [11]. Although heterogeneity has been documented between cell populations, this finding does not appear to be absolute, at least with regard to growth factors. Multiple populations of PDL cells have been shown to respond uniformly to numerous growth factors [12, 13, 14], and have also been shown to be uniform in their expression of growth factor receptors.

Though the heterogeneity between populations of these cells is well-established, there is little certainty as to the role differentiation plays in this process. In vivo, it is believed that the pluripotential stem cells from the PDL may differentiate into mineral forming cells. In vitro, recent studies have begun to document and characterize this pattern of differentiation toward a mineralizing phenotype.

Early attempts to harvest tissues from the PDL focused on two aspects. First was the mechanical separation of the PDL tissues from the root surface; and second, the need to enzymatically release the cells from the tissue explants prior to culturing. One of the earliest techniques used to culture PDL was to break apart the root of the tooth and put it in culture with the PDL tissues still adherent to the exterior of the root fragments [4]. This study was one of the first to demonstrate that PDL cells could be cultured in vitro. The only problem with this technique is that it did not preclude culturing of cells derived from the tissues in the pulpal area of the tooth. Therefore, the need to separate the PDL tissue from the root was considered. As we come to better understand specific cell functions within these unique tissues, it has become evident that this too may have its drawbacks. Most reports, both previous and current, use some method of mechanical separation of the PDL from the root surface. There are a few laboratories that are enzymatically releasing the cells from the tissue [15]. Initial work at culturing also utilized both explant techniques and cell suspension techniques with success. Early plating efficiencies were reported in the range of 0.01-0.07% [5, 16,]. Currently used techniques, as described below, are capable of providing viable PDL cell cultures on a consistent basis; however, the characterization of these cell populations and the influence of culture conditions on cell phenotypes remain incompletely understood.

2. TISSUE PROCUREMENT

The periodontal tissues surrounding the teeth can be separated into the PDL, gingival tissues, and alveolar bone. Generally, the PDL tissue used in culture explants is isolated from extracted teeth. The PDL is continuous with the gingival connective tissues at the crest of the alveolar bone. Careful attention toward anatomic characteristics that discriminate the levels of root structure contained within the bone is important in separately culturing these two tissues. Otherwise, there is potential for contamination of the PDL culture with tissues from the gingiva. By using extracted teeth to harvest PDL tissues, the bone tissue is effectively eliminated as a potential contaminating source of the tissue cultures. The extraction process usually results in the tearing of the PDL itself with a portion of the PDL remaining attached to the alveolar bone, and a portion of the PDL remaining attached to the root surface. It appears from clinical studies of both tooth avulsion / replantation, and regeneration of the attachment apparatus, that the portion of the PDL remaining on the root of an extracted tooth is critical to normal clinical healing of the attachment. Therefore, it is this tissue retained on the root surface that is cultured for in vitro investigations related to periodontal regeneration.

Once an extracted tooth has been obtained, it should be placed immediately into medium to prevent desiccation of the thin layer of tissue remaining on the root surface. Clinically, the time element with which the tissues can be maintained intact on the root surface following tooth removal, prior to successful healing with tooth replantation, has been shown to be about 30 minutes [17, 18]. Generally, we use Dulbecco's Modified Eagle's Medium (DMEM) to minimize dessication of the PDL tissue retained on the root.

The typical harvesting of PDL tissue is by mechanical scraping of the tissue remaining on the root surface following extraction. This is usually done with a sharp instrument such as a scalpel. The tissue fragments removed through this procedure are then placed into a culture system.

Before discussing culturing systems, there are several concerns that have been raised in regard to the collection of these primary tissues. These concerns include which part of the root surface is harvested, eruptive/functional status of tooth from which the tissue is collected, the age of the patient and the effect of scraping, which may remove the cemental layer along with the ligamentous tissue. In addition, a concern that has received considerable research effort is the difference between the connective tissues of the gingiva and the PDL.

It is apparent that in vitro cell cultures derived from these two distinct tissues, the gingiva and PDL, have significant differences [19, 20]. In order to ensure that tissue harvested from the root of an extracted tooth does not contain tissue fragments from the gingiva, the middle third of the root surface is utilized as the tissue source on periodontally healthy teeth. Likewise, the tissues in the apical third of the tooth are avoided to minimize the culturing of pulpal tissues. These apical tissues often have inflammatory changes associated with pulpal infection, quite often the cause for the extraction.

The concern over the type of tooth used as a source may be irrelevant. There is little evidence for differences in derived cultures according to the type of tooth. However, the potential for such differences has prompted some investigators to utilize only fully erupted and functionally active teeth.

The role of donor age appears to be of some significance, with older donor tissues being more fibrotic and less likely to lead to primary cultures [21]. The eruptive status has been considered to be a potentially important factor in primary culture characteristics. However, a recent study compared unerupted third molars with erupted premolars removed for orthodontic reasons and found no differences between the cultures established from these two sources [22].

The last factor to consider in establishing in vitro cell cultures using the mechanical isolation of tissue is the potential for the collection of differentiated cells lining the cementum surface of the tooth, cementoblasts [23]. These cells are capable of producing the mineralized outer root surface layer, and this process is continuous. The possibility exists that many of the tissue culture populations established from the PDL are actually derived from the cemental tissue. One of the more difficult aspects of explant collection is trying to separate fibroblasts derived from the PDL from cementoblasts lining the root surface. Unfortunately, there is no known marker capable of identifying these cells. At present, it is unclear which cell types are typically being cultured through this process, and one possibility is that PDL cultures demonstrating an osteoblast-like phenotype may be cultured cementoblasts.

An alternative approach to harvesting PDL has been used in a murine model. This harvesting technique is obviously impractical to use with human tissues, but is mentioned here because it presents an interesting alternative to the previous discussion. In this model, the teeth are extracted and extraction sockets allowed to heal for about one to two weeks. After this healing time, the granulation tissue within the extraction socket is removed and explanted in culture medium [24]. Earlier, it was stated that during the extraction process, the PDL tears in half and the tissue retained on the root is utilized. This model attempts to utilize the PDL tissues normally left attached to the alveolar bone. It is thought that the proliferation of cells from the remaining PDL tissue lining the extraction socket fill the extraction site during the initial healing period. It is these cells that are then harvested.

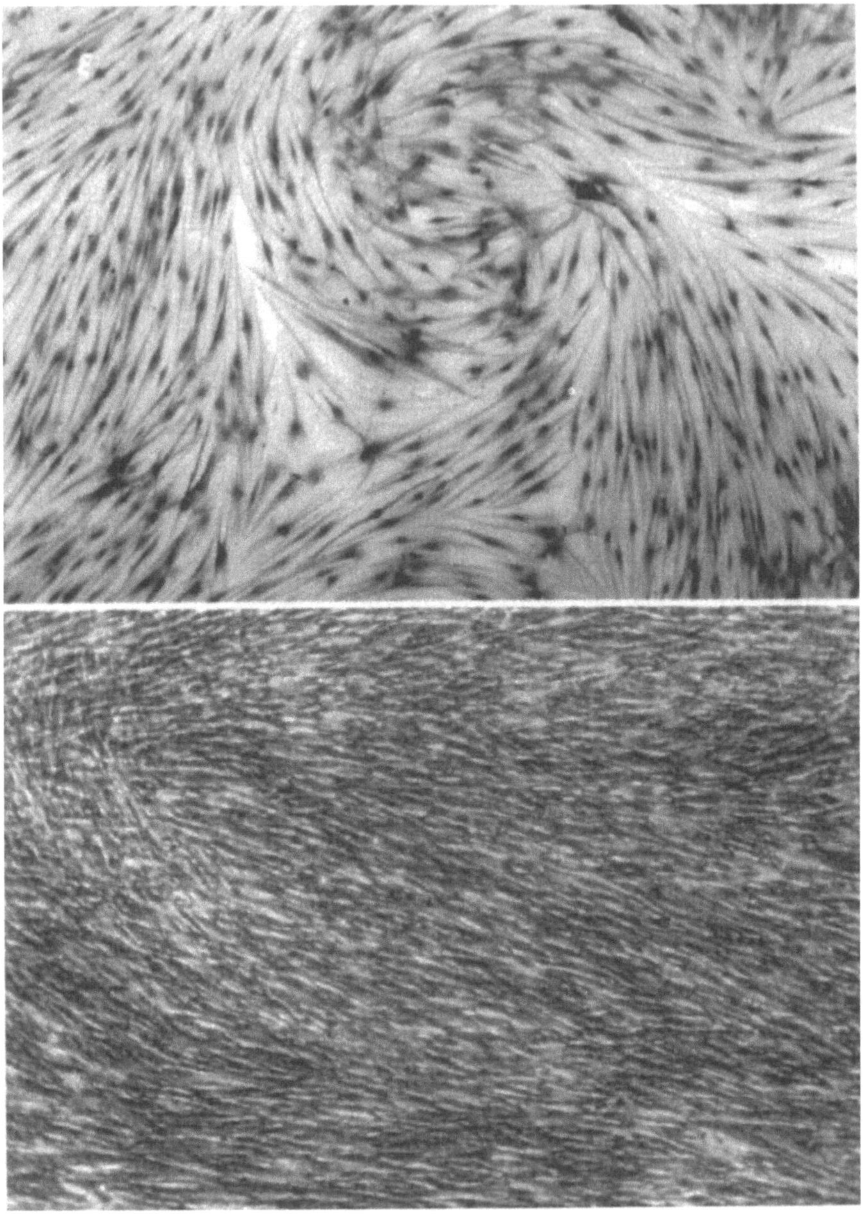

Figure 1. Phase contrast micrographs of human periodontal ligament cells showing two distinct patterns of cell growth (original magnification 10X). Top view shows whirled pattern with cell orientation in multiple directions. Bottom view shows confluent layer of cells oriented in parallel.

3. ASSAY TECHNIQUES

Human PDL cells have not been well-characterized despite considerable research efforts. This is in part due to the heterogeneity between cell cultures and between sub-populations within a given cell culture, and in part to the relatively non-distinctive nature of the cells themselves. Morphologically, the cells have a typical fibroblastic appearance that with confluence can have either a parallel or whirled pattern [10] (Figure 1). It has been suggested by one study that the differences in the patterns may correlate with the phenotype of the cells. Cells cultured from the PDL have been identified as having one of two distinct phenotypes, either fibroblastic or osteoblastic [10]. The fibroblastic cells are thought to have a parallel pattern, and the osteoblastic cells are thought to have a whirled pattern.

PDL cell populations have been well characterized for collagen production, and are becoming better characterized relative to their potential for mineralizing phenotypes. PDL cell cultures have been shown to produce collagen types I, III, and V, although there was considerable variability in percentages of production between different populations [11]. There has been one report of an attempt to develop specific antibodies for PDL. Unfortunately, the specificity of the antibodies was limited, with significant cross-reactivity to other tissues [25]. Most current research is aimed at characterizing the mineralizing phenotype that is frequently found in these cell cultures. It is thought that this phenotype correlates with the cells in vivo that are responsible for the clinical outcomes associated with regeneration of the mineralized tissues of the periodontium.

This osteogenic potential has been characterized in human PDL cultures using several markers. One of the best documented characterizations has been measurement of alkaline phosphatase activity [26]. It appears that PDL cells may have increased levels of alkaline phosphatase, particularly in cultures tending toward an osteoblastic phenotype. A second measure is the cell response to 1,25-dihydroxyvitamin D3 [27, 28]. This response has been shown to include the up-regulation of osteocalcin synthesis. Osteocalcin has been shown in humans to make up a small amount (1-2%) of the non-collagenous bone matrix, and has been identified only in bone and dentin. These findings in particular support the use of osteocalcin synthesis as an indicator of osteoblastic activity. In periodontal ligament cells, the production of osteocalcin may vary between distinct cell populations (Figure 2). In addition, the production of cAMP in response to parathyroid hormone (PTH), and the production of mineralized matrix with long-term culture, may also be used to indicate an osteoblast-like phenotype in PDL cells [29]. In addition, there are reports documenting production of other non-collagenous bone-related proteins such as bone sialoprotein and osteonectin [27].

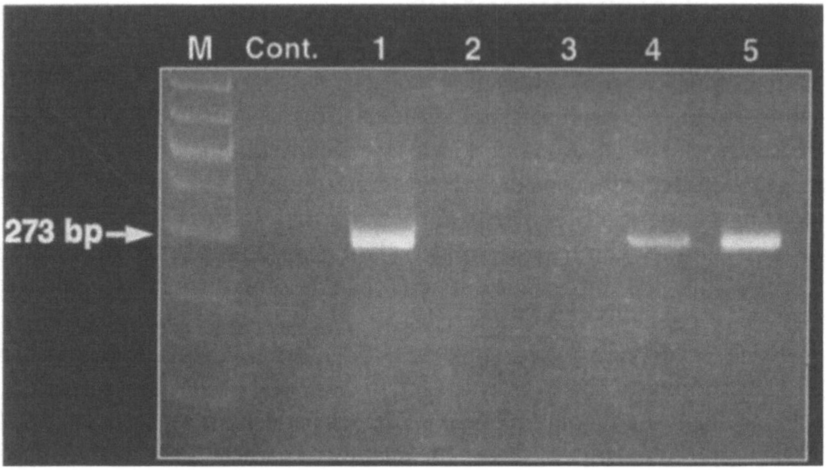

Figure 2. Variable expression of osteocalcin in human PDL cell cultures. Five PDL cultures were examined for osteocalcin mRNA utilizing RT-PCR techniques, with three of the cultures positive for osteocalcin (arrow indicating appropriately sized bands), and two of the cultures negative for osteocalcin. Amplified products for each of the five cultures are shown using ethidium bromide staining following gel electrophoresis (lanes 1-5).

In contrast to the heterogeneity found in PDL cultures, these cultures appear to be uniform in their responsiveness to growth factors. Proliferative responses to growth factors have shown these cells to be responsive to all isoforms of PDGF, and that these cells predominantly have PDGF-α receptor subunits [12, 14]. Interestingly, there appears to be less heterogeneity between cell populations or between distinct subpopulations within a given cell population with regard to this last characteristic. The responsiveness of these cells to TGF-β has been shown to be similar to that of dermal fibroblasts, including a proliferative response at low concentrations in a delayed manner, and a down-regulation of the PDGF-α receptors at higher concentrations [30, 12].

4. CULTURE TECHNIQUES

Primary PDL cell cultures are established from explants of PDL tissues as previously described. Frequently, these teeth are obtained as teeth extracted from children and young adults undergoing orthodontic treatment. Following extraction, tissue for initiating a new culture is scraped from the middle one-third of a tooth using a sharp sterile blade. Usually, only a thin layer of tissue remains attached to the root following extraction, so the collected fragments usually require no additional dissection prior to explanting.

The harvested tissue explants are placed in culture medium consisting of Dulbecco's Modified Eagle's Medium (DMEM) containing 4,500 mg/L D-glucose, 584 mg/L L-glutamine, 110 mg/ml sodium pyruvate, 4 mg/L pyridoxine hydrochloride, 3700 mg/L sodium bicarbonate, and 10% bovine calf serum (hereinafter referred to as complete DMEM). This medium is used to both establish and maintain growth of PDL cells in culture. Most studies using these cells have reported supplementing the medium with FBS, however data from our laboratory have shown that it is unnecessary to maintain these fibroblast-like cells in FBS (Figure 3). The proliferation of PDL cells cultured using bovine calf serum appears identical to that of cells cultured using FBS. After collecting the explant, the PDL fragments are placed in a drop of complete DMEM with 1% penicillin/streptomycin and 1% fungizone in a 60mm tissue culture dish by tapping the blade gently against the floor of the dish. A stainless steel wire grid is placed on top of the fragments to prevent movement of the explant and facilitate outgrowth of the new cells onto the culture surface. Some studies have described using glass cover slips or directly placing the explant onto the plastic surface for a few minutes prior to adding medium as alternatives to promote explant adhesion to the surface [20]. In our laboratory, we have found a high rate of cell proliferation is achieved from explanted PDL tissues under wire grids (Figure 4). The explant and wire grid are gently covered with four ml of complete DMEM with antibiotics. The cultures are allowed to remain undisturbed for approximately 48 hours under 37°C and 5% CO_2 to permit attachment of explants. The complete DMEM with antibiotics remains in the cultures for the first three days. Medium is then replaced every 4-6 days until sufficient cell proliferation is evident. When feeding the culture (changing the medium), care is taken not to jostle the explant fragments. Generally, cells that have migrated out of tissue fragments and attached to the wire grid can be observed within one week using an inverted microscope. Depending on the size of the explanted tissue pieces, it takes 10-14 days for the maximum number of viable cells to migrate from the tissue. In approximately 10 days, about 2/3 of the cells are clustered on the wire grid surrounding the PDL tissue, and the migration rate starts to decrease. At this time the cells are subcultured. The cells are removed from the wire grid and culture dish by brief incubation (4-5 minutes) with 1 ml 0.25% trypsin, and are then transferred to a 100 mm tissue culture dish for continued growth.

The time to confluency in 100mm plates varies with the growth rate of the culture. In most preparations, confluency is usually reached after 1-2 weeks, at which time the cultures contain about 3-4 x 10^6 viable cells. Subculture from the 100mm dish is accomplished using a brief incubation with 2 ml of 0.25% trypsin. Once the cells have been successfully removed from the 100 mm dish, the cell number is halved by adding 1ml of cells directly from the trypsin suspension to a new culture dish containing 7ml of complete DMEM. The small amount of trypsin does not appear to affect the cells. Alternatively, the trypsinized cells can be counted using a hemacytometer or cell counter and reseeded at 0.5-1x10^6 cells per 100mm plate (72,500 cells/cm^2).

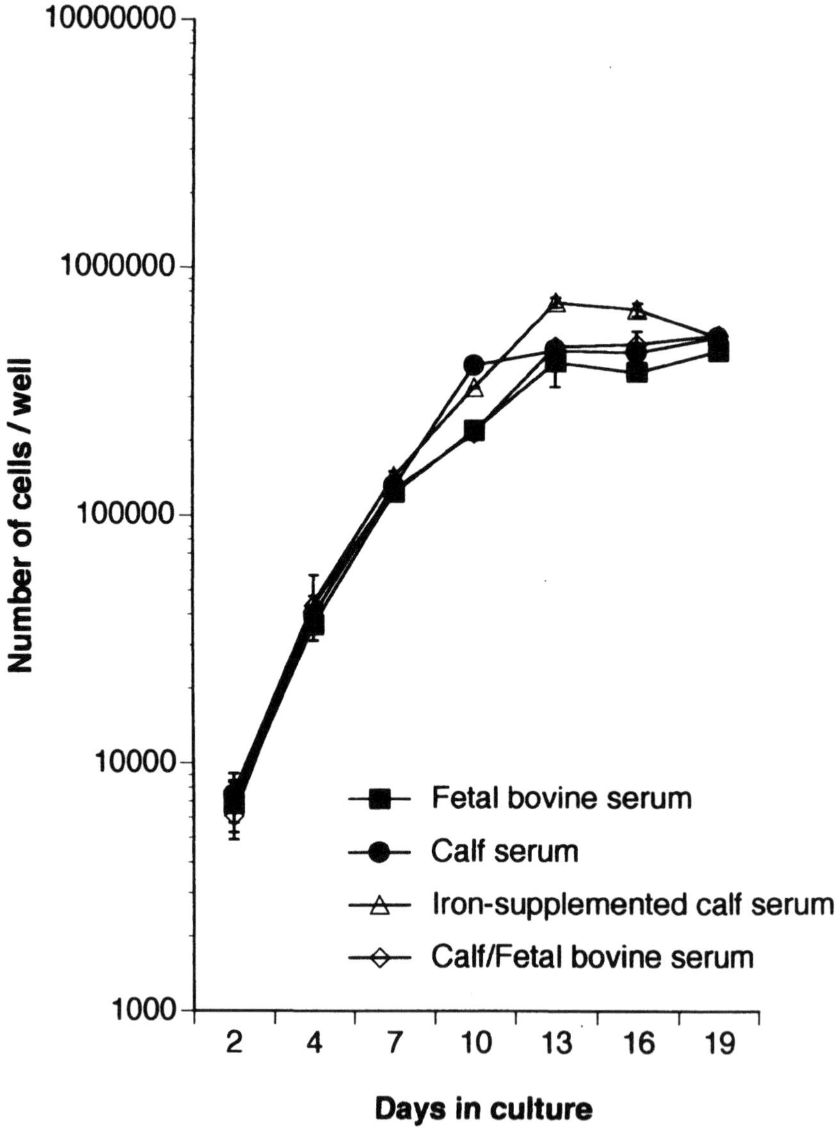

Figure 3. Effects of different culture media on cell growth. Human PDL cells were seeded in 24-well plates (5,000 cells/well) in DMEM supplemented with 10% fetal bovine serum, 10% bovine calf serum, 10% iron-supplemented bovine calf serum, or 10% combined serum containing 85% bovine calf serum and 15% fetal bovine serum. Cell numbers (quadruplicate samples) were measured after two 48-hour periods, and every three days thereafter using a Coulter counter. Data are presented as mean + standard deviation.

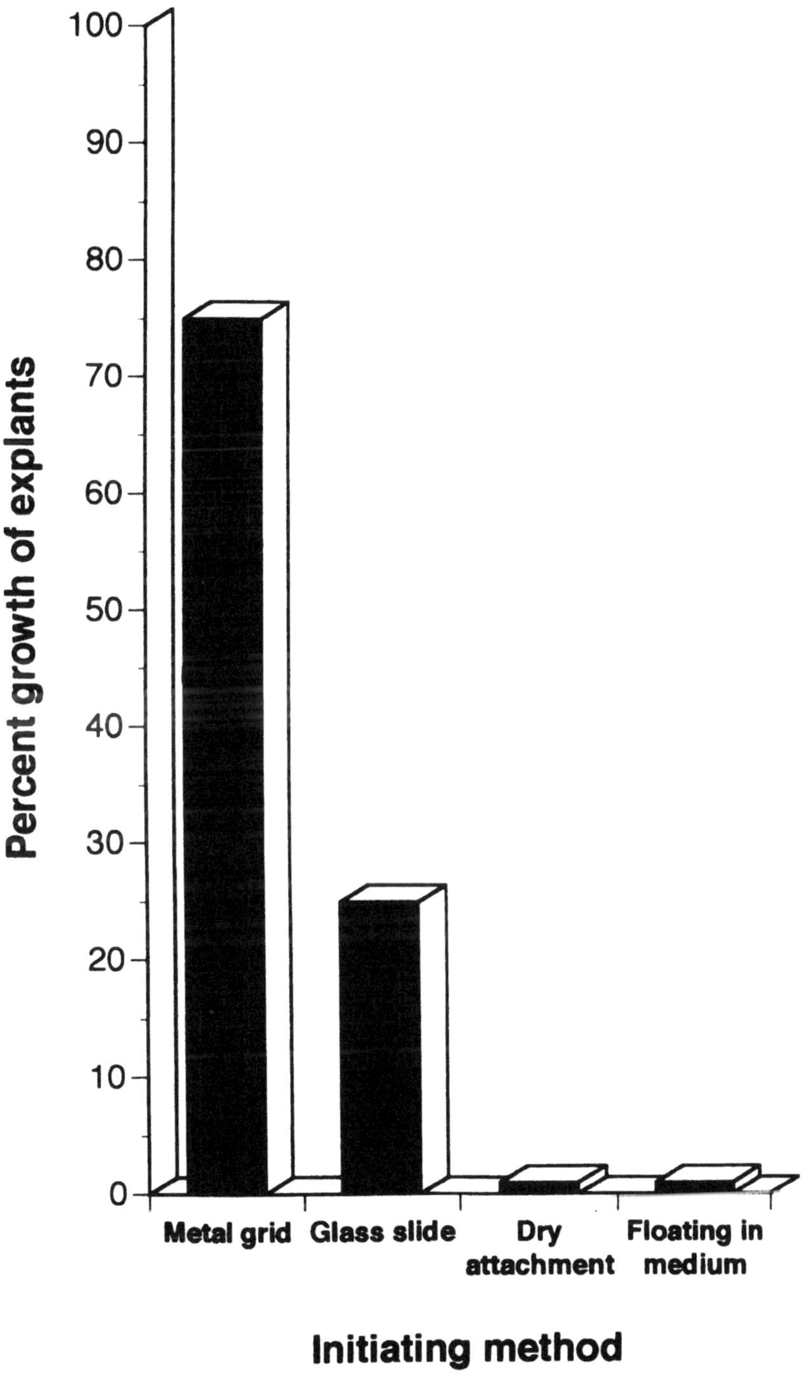

Figure 4. Comparison of four methods to initiate cell cultures. PDL tissue derived from each of four subjects was divided into four sections for culturing using one of four techniques. These techniques included the use of either a metal grid or glass slide to position the explant on the tissue culture surface, the dry attachment of the tissue to the plate surface by allowing the explant to sit on the plastic for 15 minutes prior to adding medium, or the explant left floating in the media. The data represent the percentage of explants that successfully produced

The population doubling time for successful PDL cell cultures at 37°C in DMEM containing 10% bovine calf serum is between 24 and 55 hours. Most primary PDL cell cultures grown in 10% bovine calf serum are capable of surviving 16-40 passages, or up to 150 days.

The primary PDL cultures do not retain their viability in continuous culture indefinitely. To preserve desired characteristics, cell stocks are frozen at early passage number. For the best recovery of frozen cells, subconfluent PDL cells are fed with fresh growth medium 24 hours before freezing. PDL cells are frozen at a concentration of 2-3 x 10^6 cells/ml in freezing medium (DMEM-supplemented medium containing 10% dimethyl sulfoxide and 10% bovine calf serum). Cells in freezing medium are dispensed into cryogenic vials and tightly sealed. The cells are slowly frozen by placing the cryogenic vials at 0°C for 2-3 hours, –20°C for 2-3 hours, and –80°C overnight, before transfer to liquid nitrogen for long-term storage. For successful revival of frozen cells, a vial is removed from the liquid nitrogen tank and thawed in a 37°C water bath as rapidly as possible. Then, the contents of the vial are transferred into a 100mm tissue culture plate containing 9 ml of growth medium. Cells are incubated at 37°C in humidified air containing 5% CO_2 and the medium replaced after 24 hours.

5. UTILITY OF THE SYSTEM

The unique nature of the periodontal ligament as a supporting structure and its likely role in providing progenitor tissues for the regeneration of the attachment apparatus, have created a special clinical interest in this cell type relative to both periodontal and orthodontic therapies. The growth of these cell populations in culture has allowed the investigation of biologic principles underlying periodontal wound healing. These investigations have looked at the regulatory mechanisms, such as growth factors in the wound healing response. More recently, several in vitro wound healing models have been characterized that provide the ability to address additional aspects of the regulation of wound healing. As the PDL is primary in supporting the tooth under both natural (masticatory) and artificial (orthodontic) forces, these cell cultures are also being used to address the cellular responses to mechanical stresses.

As numerous therapeutic drugs and devices are being investigated for their potential benefit in the treatment of periodontal disease, these cell cultures are also being used in the assessment of cell responses and biocompatibility. These developments are of particular interest in relation to tissue engineering. The potential for the cells to differentiate toward an osteoblastic phenotype allows investigation of materials that may be used to influence bone formation and metabolism. The development of an osteoblastic phenotype in these cells may allow for the further exploration of genetic factors that may influence the development of osteoblastic cells.

There has also been considerable research effort toward the discrimination of cells between the PDL and those from the more superficial gingival connective tissues. This research has been driven by clinical studies suggesting important differences between cells derived from these tissues in leading to the regeneration of the periodontal attachment apparatus. However, differences between these cell types have not yet been clearly defined. Finally, given the unique anatomic role of the periodontium as a protection against local oral bacteria, interactions with bacteria and bacterial products that may be significant in the development of therapeutics directed at the control of destructive periodontal disease have been studied.

6. CONCLUDING REMARKS

The PDL is an anatomically unique tissue providing functional support for the teeth in the alveolar bone. Although the characterization of the cells derived from this tissue has not provided an absolute marker, the anatomic characteristics do allow for the selection of explant materials specifically from these tissues. The high levels of destructive periodontal disease within the population, leading to the degradation of this tissue, have generated interest in the characterization of cells derived from these tissues.

The culture of these cells is relatively easy, and is made more interesting from a scientific perspective due to the apparent phenotypic dichotomy that these cultures display. It has been especially interesting that these fibroblastic cells appear to have the capacity to differentiate toward an osteoblastic phenotype in culture, thus providing an alternative model for the investigation of osteoblast development. However, a clear understanding of this developmental process remains elusive.

REFERENCES

1. Boyko G, Melcher A, and Brunette DM (1981) Formation of new periodontal ligament by periodontal ligament cells. *J Periodont Res* 16:73-88.
2. Ten Cate (1969) The development of the periodontium, in AH Melcher and WH Bowen (eds.), *The biology of the periodontium*, Academic Press, London, pp.53.
3. Gould TRL (1983) Ultrastructural characteristics of progenitor cell populations in the periodontal ligament. J Dent Res 62(8):873-876.
4. Grupe HE, Ten Cate AR, and Zander HA (1967) A histochemical and radiobiological study of in vitro and in vivo human epithelial cell rest proliferation. Archs oral Biol 12:1321-1329.

5. Brunette DM, Melcher AH, and Moe HK (1976) Culture and origin of epithelium-like and fibroblast-like cells from procine periodontal ligament explants and cell suspensions. Arch oral Biol 21:393-400.
6. Fries KM, Blieden T, Looney RJ et al (1994) Evidence of fibroblast heterogeneity and the role of fibroblast subpopulations in fibrosis. Clin Immunol Immunopathol 72:283-292.
7. Elias KA, Rossman MD, and Phillips PD (1987) Phenotypic variability among density-fractionated human lung fibroblasts. Am Rev Respir Dis 135:57-61.
8. Rose GG, Yamasaki A, Pinero GJ, and Mahan CJ (1987) Human periodontal ligament cells in vitro. J Periodontol Res 22:20-28.
9. McCulloch CAG and Bordin S (1991) Role of fibroblast subpopulations in periodontal physiology and pathology. J Periodont Res 26:144-154.
10. Piche JE, Carnes DL, and Graves DT (1989) Initial characterization of cells derived from human periodontia. J Dental Res 5:761-767.
11. Limeback H, Sodek T, and Aubin JE (1983) Variation in collagen expression by cloned periodontal ligament cells. J Periodont Res 18:242-248.
12. Oates TW, Rouse CA, and Cochran DL (1993) Mitogenic effects of growth factors on human periodontal ligament cells in vitro. J Periodontol 64:142-148.
13. Dennison DK, Vallone DR, Pinero GJ, Rittman B, and Caffesse RG (1994) Differential effect of TGF-beta 1 and PDGF on proliferation of periodontal ligament cells and gingival fibroblasts. J Periodontol 65(7):641-648.
14. Oates TW, Kose KN, Xie JF, Graves DT, Collins JM and Cochran DL (1995) Receptor binding of PDGF-AA and PDGF-BB, and the modulation of PDGF receptors by TGF-ß, in human periodontal ligament cells. J Cell Physiol 162:359-366.
15. Ragnarsson B, Carr G, and Daniel JC (1985) Isolation and growth of human periodontal ligament cells in vitro. J Dent Res 64(8):1026-1030.
16. Marmary Y, Brunette DM, and Heersche JNM (1976) Differences in vitro between cells derived from periodontal ligament and skin of Macaca irus. Archs oral Biol 21:709-716.
17. Pettiette M, Hupp J, Mesaros S, and Trope M (1997) Periodontal healing of extracted dogs' teeth air-dried for extended periods and soaked in various media. Endod Dent Traumatol 13(3):113-118.
18. Andreasen JO Borum MK, Jacobsen HL, and Andreasen FM (1995) Replantation of 400 avulsed permanent incisors. 4. Factors related to periodontal ligament healing. Endod Dent Traumatol 11(2):76-89.
19. Mariotti A and Cochran DL (1990) Characterization of fibroblasts derived from human periodontal ligament and gingiva. J Periodontol 61:103-111.
20. Somerman MJ, Archer SY, Imm GR, and Foster RA (1988) A comparative study of human periodontal ligament cells and gingival fibroblasts in vitro. J Dental Res 67:66-70.
21. Nishimura F, Terranova VP, Braithwaite M, Orman R, Ohyama H, Mineshiba J, Chou HH, Takashiba S, and Murayama Y (1997) Comparison of in vitro proliferative capacity of human periodontal ligament cells in juvenile and aged donors. Oral Dis 3:162-166.
22. Howard PS, Kucich U, Taliwal R, and Korostoff JM (1998) Mechanical forces alter extracellular matrix synthesis by human periodontal ligament fibroblasts. J Periodontal Res 33(8):500-508.
23. Grzesik WJ, Kuzentsov SA, Uzawa K, Mankani M, Robey PG, and Yamauchi M (1998) Normal human cementum-derived cells: isolation, clonal expansion, and in vitro and in vivo characterization. J Bone Miner Res 13:1547-1554.
24. Lin WL, McCulloch CA, and Cho MI (1994) Differentiation of periodontal ligament fibroblasts into osteoblasts during socket healing after tooth extraction in the rat. Anat Rec 240(4):492-506.

25. DuBois WT, Edmondson J, Milam SB, Winborn WB, Hay R, Carnes DL, Kornman KS, and Klebe RJ (1991) Monoclonal antibodies to periodontal ligament cells. J Periodontol 62:190-196.
26. Yamashita Y, Sato M, and Noguchi T (1987) Alkaline phosphatase in the periodontal ligament of the rabbit and Macaque monkey. Archs oral Biol 32(9):677-678.
27. Somerman MJ, Young MF, Foster RA, Moehring JM, Imm G, and Sauk JJ (1990) Characteristics of human periodontal ligament cells in vitro. Archs oral Biol 35(3):241-247.
28. Basdra EK and Komposch G (1997) Osteoblast-like properties of human periodontal ligament cells: an in vitro analysis. Eur J Orthod 19:615-621.
29. Arceo N, Sauk JJ, Moehring J, Foster RA, and Somerman MJ (1991) Human periodontal cells initiate mineral-like nodules in vitro. J Periodontol 62(8):499-503.
30. Battegay EJ, Raines, Seifert RA, Bowen-Pope DF, and Ross R (1990) TGF-ß induces bimodal proliferation of connective tissue cells via complex control of an autocrine loop. *Cell* 63:515-524.

Chapter 4

Vascular Smooth Muscle

Diane Proudfoot and Catherine M Shanahan
Department of Medicine, University of Cambridge, ACCI, Level 6, Box 110, Addenbrooke's Hospital, Hills Road, Cambridge CB2 2QQ, U.K. Tel/Fax: 0044-1223-331504; Email: dp or cs131@mole.bio.cam.ac.uk

1. INTRODUCTION

Vascular smooth muscle cells (VSMCs) are a major component of arteries, veins and small blood vessels. Their main function is to maintain vascular tone, via co-ordinated vasoconstriction and vasorelaxation, thereby regulating blood pressure and flow. In addition to this contractile function, VSMCs are the only cells in the vessel wall capable of migration, proliferation and extra-cellular matrix production; properties necessary for vascular repair at sites of injury. Thus, VSMCs play an essential role in a number of vascular pathologies including atherosclerosis, hypertension, re-stenosis after coronary angioplasty and bypass grafting and small artery disease associated with rejection after heart transplantation. Because of their association with these common pathologies, a vast number of researchers have studied their properties. Historically, most studies on VSMCs have been carried out using cells from non-human sources. Studies, particularly of rodent cells, have demonstrated that VSMCs are extremely plastic and can display a number of different morphologies and gene expression profiles related to different cell functions. This 'phenotypic diversity' is most obvious during vascular development and is reiterated in vascular disease. Additionally, avian models have shown that VSMCs of different vascular beds may be intrinsically different, as they are recruited in development from at least two different sources: mesenchyme and neural crest. In this chapter, we provide some background knowledge of VSMCs *in vivo* and *in vitro*. Importantly, we highlight the more recent observations and advances in knowledge of VSMC function, particularly in disease, which have been obtained from *in vitro* studies of human VSMCs.

M.R. Koller, B.O. Palsson and J.R.W. Masters (eds.), Human Cell Culture Vol V, 43–64.

1.1 Location of VSMCs in the blood vessel wall

The vessel wall is made up of three distinct layers which vary slightly in composition between individual arteries/veins. The layer which is exposed to the circulating blood is the tunica intima, which is covered on the lumenal side by a single flattened sheet of endothelial cells lying on basement membrane of collagen type IV. In some arteries, such as the aorta and coronary arteries, there are also VSMCs in the intima. The role of the intima is to provide a smooth, non-thrombogenic surface and to act as a permeability barrier to cells and macromolecules. Beneath the intima is the internal elastic lamina, which is a sheet of fenestrated elastic fibres. The cell layer beneath this is the tunica media which contains longitudinal, spirally arranged SMCs interspersed with elastin fibers and an extracellular matrix rich in collagen I, fibronectin, and proteoglycans. The VSMCs are surrounded by a basement membrane containing collagen IV, laminin and heparan sulfate [1]. The outermost layer is the tunica adventitia, which is separated from the media by the external elastic lamina, and contains fibroblasts surrounded by collagen, proteoglycan matrix, vasa vasorum and innervation [2] (see Figure 1).

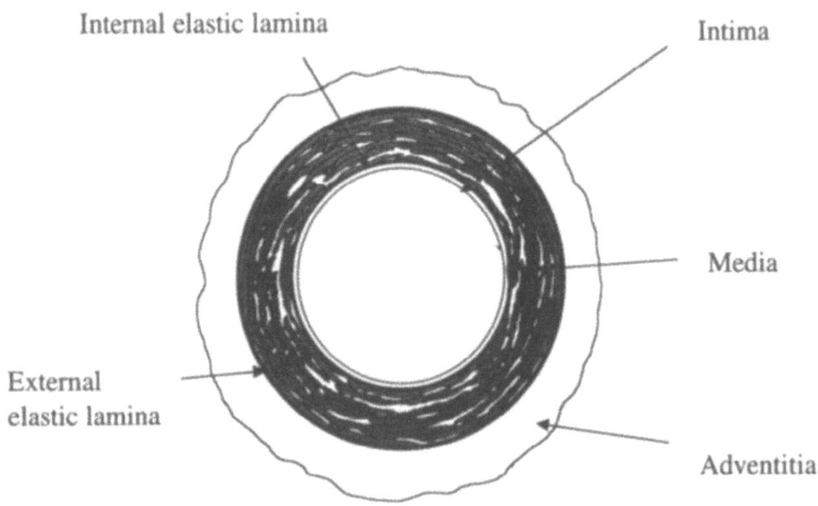

Figure 1. The blood vessel wall is made up of three distinct layers which vary slightly in composition between different arteries and veins; the tunica intima, the tunica media (containing VSMCs), and the tunica adventitia. These layers are separated by the internal and the external elastic lamina (see section 1.1).

1.2 VSMC phenotype and importance in disease

In the normal vessel media the VSMCs exist in a contractile phenotype. However, VSMCs are not terminally differentiated and in response to vascular injury, particularly in atherosclerosis, VSMCs de-differentiate, migrate and proliferate, leading to their accumulation in the vessel intima. Thus, the VSMCs which occupy the media are phenotypically distinct from their intimal counterparts [3, 4]. In general, intimal VSMCs have lower levels of contractile proteins, contain fewer myofilaments, more synthetic organelles (e.g. sarcoplasmic reticulum and mitochondria) and most importantly, express a variety of gene markers not normally expressed in the media [5-8]. This property of VSMCs to undergo phenotypic change was considered as a key pathogenic factor by Ross and Glomset in 1976 when they proposed the 'response to injury' hypothesis for the development of atherosclerosis [9, 10]. In atherosclerosis, the most widely studied of vascular diseases, VSMCs are found in the enlarged intima in differing quantities, depending on the type and stage of the atherosclerotic lesion. Historically, VSMCs in atherosclerosis were considered to be detrimental as their presence in the intima led to a narrowing of the lumen. However, more recently it has been suggested that the stability of the lesions, that is their propensity to rupture and initiate a broad range of clinical symptoms, is dependent on the VSMCs within the lesion. Stable plaques contain VSMCs and matrix, which form a thick fibrous cap above a lipid core. In contrast, plaques which are unstable and more likely to rupture contain few VSMCs, many inflammatory cells (most of which are macrophages) and a thin cap [11]. Therefore, although intimal VSMCs are associated with vascular disease, they may be beneficial in some circumstances in that they add stability to plaques by forming a protective cap rich in extracellular matrix. Furthermore, more recent studies have suggested that VSMCs may also contribute to lipid accumulation, calcification and cell death within the plaque. Therefore, understanding the mechanisms which regulate VSMC phenotypic change in association with specific vascular diseases will be crucial for our ability to modulate disease processes. Studies to date suggest that VSMC phenotypic modulation occurs in response to a vast array of environmental stimuli. The elaboration of human tissue culture models of the VSMC phenotypes exhibited in vascular disease will enable us to further dissect specific functions and regulatory mechanisms.

1.3 VSMC heterogeneity and tissue culture models

When VSMCs from the normal medial layer are cultured *in vitro*, they undergo phenotypic change and de-differentiate and proliferate to become 'synthetic' cells. These 'synthetic' cells have broad similarities to many intimal cells and have therefore been extensively used to model disease-associated VSMCs. However, recent evidence from studies of human vascular disease *in vivo* has clearly demonstrated that the definition of VSMCs as either 'contractile' or 'synthetic' is simplistic. Furthermore, using animal tissue culture systems, particularly rat and bovine VSMCs, it is possible to model multiple VSMC phenotypes *in vitro* [12, 13]. In adult bovine arteries, four distinct VSMC phenotypes can be recognised. These originate in different layers of the media and sub-intima. Cells cultured from specific layers retain distinct phenotypes as judged by morphology, smooth muscle marker expression and responsiveness to growth stimuli. Frid et al. concluded that these different cell types exist in the media to perform different functions (i.e. contraction or repair) [12]. Similarly, cloning of VSMCs from rat arteries identifies at least two distinct phenotypes; elongate and cobblestone, with the latter being almost exclusively present in the neointima after balloon injury-induced damage [13]. These cells are also more prevalent in cultures derived from neonatal vessels compared with adult vessels. Analysis of gene/protein expression in these cell cultures has led to the identification of specific gene markers such as cellular retinol-binding protein-1, which has distinct patterns of expression for each phenotype [14].

Although VSMC heterogeneity has also been observed in human blood vessels *in vivo* [15, 16], particularly between VSMCs occupying the sub-intimal layer and different layers within the media, tissue culture models of these differences are not as well characterized as those in other species. *In vitro*, human VSMCs can display different characteristics, including different cell shapes, cell sizes, protein expression and growth patterns [17]. For example, human VSMCs isolated from young arteries contain populations of cells with higher proliferative rates than VSMCs from adult arteries [18]. In addition, in a human VSMC culture model where 'contractile' VSMCs were established within collagen gels and compared to proliferating VSMCs, the cell surface protein CD9 was expressed in greater amounts in proliferative cells [19]. Furthermore, Bennett et al. found that human VSMCs derived from the vessel intima exhibit a higher natural rate of apoptosis in culture than cells derived from the normal vessel media [20]. Our group has also found that human arterial VSMCs cultured from normal media can exist in different forms judged by morphology, and can calcify *in vitro* (Figure 2, described in section 4, and ref 21). However, many of these morphologies do not appear to be stable in culture and specific gene markers of different human VSMC sub-populations *in vitro* have yet to be identified.

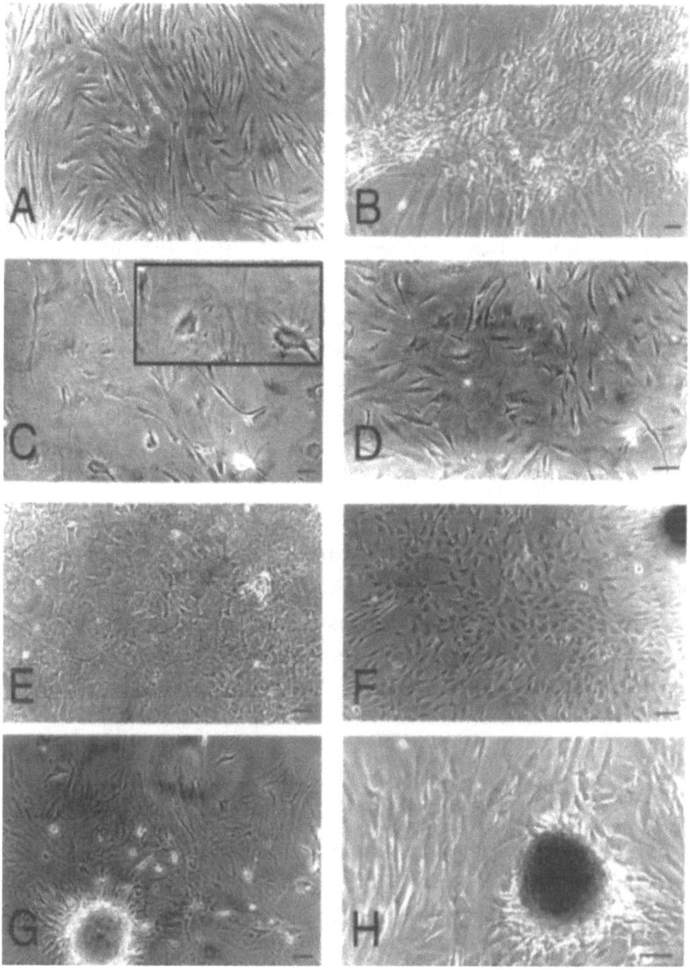

Figure 2. Human VSMC morphology. A. Typical spindle shaped VSMCs derived from aortic medial explants. B. Classic 'hill and valley' post-confluent growth of spindle shaped cells derived from aortic medial explants. C. Large, rounded senescent-type cells, some with long extending processes, derived from aortic medial explants. The insert is an enlargement of part of this culture which shows the morphology of this culture in greater detail. D. Mixed population of cells appearing as either spindle, elongate, large rounded or stellate. These cells were obtained by enzymatic dispersion of the aortic media. E. Mostly contact-inhibited morphology forming a cobblestone-like layer of VSMCs derived from aortic medial explants. If these VSMCs are allowed to remain as a monolayer, they gradually curl up as a sheet of cells and are difficult to trypsinize. Fugita et al (18) described this type of VSMC morphology as similar to neonatal VSMCs. F. Mostly small, highly proliferative cells derived from an aortic medial explant (shown top right). G. Pericytes derived from placenta microvessels with a multi-cellular nodule. Note the prominent actin filaments in these cells. H. VSMCs, mostly of the spindle-shape, with a large multi-cellular nodule derived from aortic medial explants. (The bar in each figure represents approximately 10μm).

1.4 Importance of using human VSMC cultures

It is most appropriate to use human VSMCs whenever possible. Interpretation of data from rodent cultures and application to human VSMC disease can lead to problems as animal models do not adequately mimic human disease, particularly atherosclerosis. For example, there have been many studies where inhibitors of re-stenosis have been identified in the rat. However, these same substances proved ineffective in human clinical trials. Furthermore, studies have demonstrated major differences in gene expression between human and rodent VSMCs [22, 23]. Rat neonatal aortic VSMCs express high levels of the multifunctional matrix protein osteopontin, but this is not true for human VSMCs in culture, where RT-PCR is required to detect this protein [23]. Additionally, cultured human VSMCs express very low levels of the bone cell marker, alkaline phosphatase, unlike bovine VSMCs [24].

Although most *in vitro* experimentation is still carried out on VSMCs from non-human sources, cultures of human VSMCs have been described in the literature since the late 1950s [25]. The use of human VSMCs largely depends on the availability of human tissue and success in growing cultures. In the following sections, we concentrate on the methods for isolation and culture of VSMCs from a number of different vascular sources in man.

2. TISSUE PROCUREMENT AND PROCESSING

2.1 Obtaining human blood vessels

VSMCs derived from different vessels may exhibit different properties *in vitro*. Therefore knowing the exact vascular source of any material being used is important. The aorta is made up of VSMCs that originate from neural crest in the aortic arch and from mesenchyme in the thoracic segment, whereas coronary artery VSMCs may be derived from a different embryonic source [26, 27]. However, the cells used in any laboratory will normally be limited by local availability. Arteries used for the isolation of human VSMCs include; thoracic aorta, aortic arch, abdominal aorta, coronary artery, carotid artery, saphenous vein, pulmonary artery, umbilical artery, renal artery, tibial artery and femoral artery. The aortic samples are obtained from organ donors. Carotid arteries are obtained from patients undergoing carotid endarterectomy and saphenous veins from patients having bypass graft surgery. Our group has obtained leg vessels such as tibial and femoral arteries in collaboration with surgeons performing amputations. We have also obtained VSMCs from the micro-vasculature using microvessel-rich human placenta as a source. This tissue is readily available and can also be used as a source of umbilical artery/vein VSMCs. Some groups have reported using neonatal human arterial VSMCs [18], and it is possible to obtain postmortem material [17, 18]. However, in our hands, this material cannot be obtained quickly enough, is often contaminated with micro-organisms, and cannot be cultured successfully.

We would recommend using material freshly obtained at surgery. Typically after surgery the vessel is placed in serum-free culture medium and kept at 4°C. Ideally the vessel should not be stored for more than 24 hours before use. The shorter the time from excision to culture, the greater the chances that the material will produce viable cultures. Timing, which is crucial, depends on communication between transplant/surgical teams and the researcher.

2.2 Processing tissue for VSMC isolation

2.2.1 Explant method

Our experience in culturing VSMCs is mainly from human aorta, although the methods used for this vessel can be applied to others. Explant culture of medial tissue is the simplest and most reliable method. The vessel is removed from the container and washed in serum-free medium. It is then cut open longitudinally using forceps and scissors. The lumenal side of the vessel is freed of endothelium by lightly scraping the surface with a scalpel blade and again washing the surface with serum-free medium. If the vessel is diseased and contains fatty streaks or atherosclerotic plaques, these should be avoided and removed so that VSMCs are cultured from the normal media. The media which contains layers of VSMCs is now exposed, and these can be removed easily in strips by peeling away the layers using forceps. The external elastic lamina separates the last layer of VSMCs from the adventitia, providing an easily recognizable physical barrier between VSMCs and fibroblasts. The medial layers containing VSMCs are placed in a separate petri dish and cut very gently using a scalpel blade to produce 1mm x 1mm tissue pieces. It is important that the tissue is not heavily sheared by the scalpel as this can affect culture efficiency. To reduce damage to exceptionally large samples, a mechanical tissue chopper can be used to generate small tissue explants. When the tissue pieces have been generated, they are placed in 6-well plates. It is essential that the explants adhere to the tissue culture plastic to ensure VSMC migration. To achieve this, either a sterile coverslip is placed on top of the explants and covered with medium, or the explants are given only a small volume of medium for the first 24 hours of culture. Cells should start to grow out radially from the explants within 2-3 weeks.

It is noteworthy that some groups have reported that different layers within the aortic media contain different types of VSMCs [12, 16]. Additionally, we have cultured intimal versus medial VSMCs and found some differences in growth patterns in culture. It may therefore be appropriate in some instances to culture different layers of the intima/media separately.

2.2.2 Enzyme dispersion method

Enzyme dispersion of the media is not used as a standard method for the isolation of human aortic VSMCs because the yield of cells is lower than that of the explant method. However, the advantage of enzyme dispersion is that the whole population of cells from the media is represented, rather than selected cells that are capable of migrating from the tissue. In addition, the cells isolated by enzyme dispersion may represent a phenotype more closely associated with those in the vessel wall since the time to isolation is much quicker (24 hours as opposed to 2-3 weeks). Therefore, some researchers may wish to use this method for their application. It should be noted that this method can also be used to successfully isolate mRNA from the medial layer of VSMCs [21].

Before enzyme dispersion, the vessel sample is prepared to the stage of tissue chopping, as described above for explants. The tissue pieces are then placed in a sterile conical flask and incubated in serum-free medium containing 3 mg/ml collagenase and 1mg/ml elastase (both from Sigma). This incubation is carried out in a shaking waterbath overnight at 37°C with occasional dispersion using a wide-mouthed pipette. If the tissue pieces are not completely digested overnight, the cells which have dispersed are removed by centrifugation (1000 rpm, 5 minutes) or by sieving through a 70 μm filter (Falcon), and the remainder left to disperse further. Ideally the dispersed cells should be collected over a time course to maximise the yield of viable cells. When dealing with a blood vessel which is much smaller than an aorta, such as a piece of coronary artery, the incubation in collagenase and elastase should be carried out for much shorter times. Longer exposure to the collagenase/elastase medium may harm the cells. The dispersed cells should be centrifuged and resuspended in fresh medium before plating in culture dishes at approximately 8000 cells/cm^2. Cells isolated from small arteries will be few in number and should therefore be plated in small wells such as 24- or 96-well plates to ensure an adequate cell density. Because human VSMCs require a close association with other cells to maintain survival and growth, we have not been successful in single cell cloning by limiting dilution of human VSMCs.

2.3 Isolation of VSMCs from the micro-vasculature (pericytes)

Isolation of VSMCs from very small vessels first requires the isolation of microvessels from the surrounding tissue. We have isolated micro-vascular SMCs (or pericytes) from human placenta, although others have used the brain as a source [28]. Because of the ease of obtaining fresh microvessel-rich human placentas from maternity units, pericytes from this tissue are relatively easy to isolate. First, a central section of the fresh placenta which is rich in villi, distant from large blood vessels and the outer membrane, is dissected and washed thoroughly in serum-free medium. The tissue is then manually chopped into small pieces and placed in serum-free medium containing 3 mg/ml collagenase. After incubation at 37°C in a shaking water bath for 3 hours, the digested material contains single cells and microvessels. To separate the microvessels from the cells in suspension, the material is sieved through a 70 μm filter (Falcon) and washed with serum-free medium. The microvessels, which remain on the sieve, can then be removed and placed into 25 cm^2 flasks in medium containing 20% serum. After 5-6 days in culture, pericytes as well as endothelial cells grow out from the adherent microvessels. Without special supplements for endothelial cell growth, the endothelial cells die after two rounds of trypsinization/passaging, leaving pure pericyte cultures. Alternatively, endothelial cells can be selectively destroyed from mixed cultures of endothelial cells and pericytes by plating established cell cultures onto dishes coated with l-leucine methyl ester. Due to differences in esterase expression, this substance is toxic to endothelial cells but not to pericytes [29].

2.4 Isolation of VSMCs from diseased vessels

The isolation of VSMCs from atherosclerotic tissue is complicated by the presence of several other cell types, mainly macrophages and lymphocytes, although other inflammatory cells may also be present. Depending on the type of lesion, VSMCs may be abundant or scarce. In addition, successful isolation of VSMCs from diseased vessels is not guaranteed. Explant culture of VSMCs from carotid endarterectomy specimens is possible. Although a number of cell types may initially grow from the explanted material, only VSMCs will survive beyond primary culture into the first passage. Alternatively, if cell dispersion is used, undesired cell types can be removed from collagenase-digested suspensions before plating by using magnetic beads with specific cell surface ligands (Dynal). Isolation of VSMCs from re-stenotic tissue should be somewhat easier than isolation from atherosclerotic tissue because the intima of re-stenotic vessels contains mainly VSMCs. However, specimens from coronary atherectomies of re-stenotic vessels are extremely small, typically 1 mm thick and up to 10 mm long. In our hands, we find that this cylinder of tissue is too small for preparation of explants. Successful isolates of VSMCs can be obtained by digesting the material in 3 mg/ml collagenase and 1mg/ml elastase for 1 hour at 37°C in a shaking waterbath and plating cells in a few wells of a 96-well plate. Dartsch et al. reported that plaque atherectomy specimens could be used to generate explants, although enzyme dispersion was also used [30]. Both methods generated only small numbers of cells, approximately 30,000 cells per 100 mg tissue.

A major problem with diseased tissue is the potential contamination by cell types similar to VSMCs which may be present in the plaque, such as mesenchymal cells, pericyte-like and fibroblast-like cells. Assays for VSMCs are therefore very important and are described in section 4.

3. CULTURE TECHNIQUES

3.1 Media and supplements

The media which have been used for human VSMC culture include Medium 199 and DMEM buffered with 3.7 g/L $NaHCO_3$ and 5% CO_2. Antibiotic supplements recommended are 100 IU/ml penicillin, 100 μg/ml streptomycin and 250 ng/ml amphotericin B (the latter is important to avoid fungal contamination in the early stages of culture). The medium is also supplemented with fresh 4 mmol/l of L-glutamine. The sera used for supplementing the medium commonly derive from fetal bovine sources, although some researchers may be able to use human or autologous serum if the VSMCs are derived from a surgically removed artery, for example during bypass grafting. Serum is used at concentrations of either 10% (for passaged cells) or 20% (for explant cultures) and should be heat-inactivated at 56°C to destroy complement. Since the concentration of some components of FBS differ drastically between different batches, serum testing is essential. Growth factors are not commonly required in addition to serum. A limited number of serum-free systems have been tested where it is necessary to use defined supplements, including endothelial cell growth factor (ECGF), to maintain cell viability [17]. However, cultures must first be established in the presence of serum, which is then gradually replaced with serum-free medium.

3.2 Culture vessels

No specific brand or type of flask is necessary for human VSMC culture. We recommend using Falcon 6-well plates for the adherence of media explants and Corning 25, 75 or 150 cm^2 flasks for the passaging of confluent cultures. Human VSMCs also adhere and grow well on tissue culture-treated glass slides and tissue culture inserts for co-culture experiments. Culture dishes can be pre-coated with extracellular matrix proteins to test the effects of attachment factors on cell adhesion and proliferation. Boyden chambers have been used to culture human VSMCs for measuring their migration in response to various factors [31], and specialised devices have been created for measuring mechanical strain on human VSMC cultures [32]. VSMCs can also be cultured in flexible dishes (FlexCell International) to investigate the effects of mechanical forces.

3.3 Passaging of cultures

With feeding every 3-4 days, VSMCs start to migrate from adherent explants within 2-3 weeks of culture. When the migrated cells have formed confluent areas, the cells are trypsinized by washing gently three times in calcium-free balanced salt solution, then incubating in freshly thawed trypsin (0.5 mg/ml)/EDTA (0.2 mg/ml) solution at 37°C (ideally in the tissue culture room incubator). After 3 to 5 minutes the cells begin to 'round up,' and if the plates are tapped lightly the cells are released into the medium. Fresh medium containing serum (approximately 5 ml for a T75 flask) is then added to the wells to inactivate the trypsin. If coverslips have been used to encourage adherence of explants to the culture dish, it is likely that some cells will remain on the coverslip. To remove these additional cells the coverslip can be inverted using sterile forceps, and then the washing and trypsinization steps are repeated. The cells, which are now in suspension in serum-containing medium, can be pooled and placed in a larger culture vessel. It is highly recommended that cells are plated out relatively densely, covering about 50% of the culture dish (i.e. passaging 1:2 v/v). This density encourages the growth of human VSMCs and is also recommended for pericytes. For maintenance of cultures, it is recommended that the medium is changed every two days. The typical passage survival time for normal medial VSMCs is up to ten passages, although pericyte cultures do not often survive beyond passage five. For VSMCs isolated from diseased vessels, the passage survival time is much lower and is typically up to three passages [33]. Senescence of VSMCs may be accompanied by the production of a large amount of cell debris.

Human VSMCs and pericytes can be stored frozen in 10% DMSO and 20% FBS. Since DMSO inhibits VSMC growth, cells should be thawed quickly and either centrifuged at 1000 rpm for 5 minutes to remove the DMSO or the medium changed as soon as the cells have adhered to the culture flask.

4. ASSAY TECHNIQUES

4.1 Identification of VSMCs

Because of the heterogeneity and adaptive modulation displayed by VSMCs in culture, a number of criteria must be used to ascertain that cultured cells are derived from VSMCs. This is particularly important for cells derived from atherosclerotic vessels, as some cell types such as myofibroblasts and primitive mesenchymal cells share expression of a number of SM-contractile proteins with VSMCs. Indeed, there are no definitive markers to differentiate between these cell types and VSMCs. Therefore, in order to determine that a particular culture is VSMC-derived, the researcher must use a number of criteria including origin of the cells, morphology and expression of a 'battery' of smooth muscle markers. To date, many of these markers have been derived from studies in other animal species, and there is need for further identification of more appropriate and specific markers of human VSMC phenotypes *in vitro*.

4.1.1 Morphology

Cultured human VSMCs are morphologically heterogeneous. Typically, as VSMCs approach confluence, they exhibit a 'hills and valleys' morphology where cells make contacts with each other and retract. However, VSMCs can also appear as elongate, spindle, cobblestone or large, rounded senescent cells. Examples of these various morphologies are shown in Figure 2. As mentioned previously, these types of VSMCs are not necessarily stable in culture, and we have found that the small, primary cells shown in Figure 2F can change their morphological appearance after passaging. Occasionally, it has also been noted that some cells migrating from explants contain lipid droplets which appear as small cytoplasmic vesicles (using phase-contrast microscopy) and stain positively with Oil-Red O. In addition, electron microscopy can be used to analyze features of VSMCs; cells with a more 'contractile' phenotype appearing as a cell with a heterochromatic nucleus and abundant actin and myosin filaments, compared with 'synthetic' VSMCs with a euchromatic nucleus, prominent sarcoplasmic reticulum and Golgi complex (34).

4.1.2 Multicellular nodule formation by VSMCs

The capacity for human VSMCs to form 3D nodules or spheroids at post-confluence (as shown in Figure 2) has been well described for arterial VSMCs [17, 21, 35, 36] and pericytes [37]. We have found that human arterial VSMCs (aortic, tibial and femoral) and human placental pericytes form nodules in culture which calcify after approximately 30 days [21, 38]. This property of VSMCs exists in almost all isolates that we have prepared. The only morphological-type which does not form nodules are the 'cobblestone' cells which appear to be contact-inhibited. Although not unique to VSMCs (mesenchymal cells with 'osteoblastic properties' may also form nodules), this property will distinguish VSMCs from fibroblasts.

4.1.3 VSMC contraction

A unique property of VSMCs compared with other cell types present in blood vessels is contractility. However, measuring contraction in cultured cells is not straightforward, as VSMCs lose many of their contractile properties with time in culture. Therefore, most studies of VSMC contraction involve culture of blood vessel ring segments that retain some contractile responses, and these can be measured by treatment with agonists such as noradrenaline or KCl and subsequent measurement of contractile force and/or intracellular free calcium concentration. However, in a study by Shirinsky et al., contraction of enzyme-dispersed rabbit VSMCs occurred in response to histamine [39]. Contraction was observed during the first six days of culture and was measured optically. Measurement of aspects of VSMC contraction in culture is therefore feasible. More recently, contraction of individual human VSMCs in culture has been reported [39a].

4.2 Cell markers

4.2.1 Antibodies

α-smooth muscle (αSM) actin has been universally used as a marker for VSMCs [40]. However, some studies have reported that endothelial cells and adventitial fibroblasts can express α-SM actin under certain culture conditions [41, 42]. Therefore, a battery of SM markers must be used to identify cells *in vitro*. Most VSMC markers are contractile proteins, and antibodies for many of these are widely available commercially. Some common markers include; smooth muscle myosin heavy chain (SMMHC), calponin, SM22α, desmin, h-caldesmon, metavinculin and smoothelin. SMMHC is the most specific marker. However, expression of this protein is not always maintained in SM cultures. In general, to confirm that cultured cells derived from the vessel media are VSMCs, we require positivity for α-SM actin and a least one other SM contractile protein (e.g. calponin or SM22α). Cells are tested by immunohistochemistry, which requires very few cells grown in multi-well chamber slides, fixed in paraformaldehyde. Alternatively, Western analysis can be performed on cell lysates.

It may also be appropriate to confirm that cultures are von Willebrand factor negative if there is concern about endothelial cell contamination (for example, when isolating VSMCs from small blood vessels). Separate markers have also been described for microvascular SMCs (pericytes) such as 3G5 [43], but there is debate about the specificity of this marker [44]. The high molecular weight melanoma-associated antigen is also a marker for pericytes *in vivo* [28].

4.2.2 Molecular Profile

Gene expression studies are probably the most reliable method for determining the type of cell that has been cultured (although they generally cannot detect contamination of a culture with another cell type). mRNA can be extracted from cultures and gene expression can be analysed by either Northern or RT-PCR. The most reliable markers for analysis are again the SM-specific contractile proteins including α-SM actin, SM22α and calponin, which are easily detectable on Northern analysis. SMMHC may be detectable by Northern analysis, but RT-PCR may be necessary. To date, for human cells, there are no markers that will definitively identify a VSMC. However, expression of a number of SM contractile proteins detectable by Northern analysis is sufficient, particularly if the anatomical source of the material is certain [45].

4.2.3 Extracellular matrix

As described in the introduction, VSMCs *in vivo* are surrounded by various extracellular matrix components. Extensive studies of the types of matrix secreted by human VSMCs *in vitro* have not yet been performed. In other species, VSMCs secrete various collagen types, fibronectin, laminin, proteoglycans, elastin, and SPARC, as well as integrin receptors for binding to extracellular matrix (reviewed in 34). Human VSMCs synthesize matrix degrading enzymes as well as endogenous inhibitors of matrix degradation [46, 47]. Although these extracellular matrix proteins are not unique to VSMCs, the expression in culture of a large subset of these matrix components, combined with expression of a SM contractile marker, may be indicative that the cells are SM in origin.

4.3 Growth

The growth of human VSMCs in culture is much slower than that of VSMCs from other species. Typical cell cycle times for human VSMCs are approximately 44 hours (although they may be as long as 72 hours), whereas rat VSMCs double within 35 hours [48]. The cell cycle can be synchronized in human VSMCs after quiescence for 72 hours in medium such as M199 supplemented with 0.1% serum. The cell cycle time can then be measured using standard methodologies. The cell cycle time for human VSMCs varies between isolates derived from the normal media, and tends to become longer when thc cells approach senescence and become large and rounded. Additionally, the origin of the VSMCs may introduce variability, with cells from primary atherosclerotic lesions having different proliferative capacity from cells from re-stenosing lesions [30]. In Dartsch's study [30], the primary lesion VSMCs were enlarged, highly adhesive, fibroblast-like cells, which grew more slowly than the smaller, less adhesive cells isolated from re-stenotic lesions. However, confirmation is required to determine if these features are consistently different in studies from a number of laboratories.

5. UTILITY OF SYSTEM

5.1 VSMC growth and migration

VSMCs in culture have been widely used as a model of VSMC proliferation, migration and differentiation in the artery wall [34, 49]. Analysis of the regulators of VSMC growth and migration provide insights into the role of VSMCs in the development of atherosclerosis, re-stenosis and hypertension. Most studies however, have used animal cells in culture and animal models of balloon injury and subsequent repair. Consequently, many drugs have been discovered that inhibit VSMC migration and proliferation in animal re-stenosis models, such as rat carotid artery balloon injury, but most have been unsuccessful in human trials. These studies in particular have highlighted the necessity for the generation of human tissue culture models of functional VSMC phenotypes. Human VSMC cultures are therefore of great use in preliminary drug analysis for re-stenosis research. Preliminary evaluation of a drug in monolayer culture, followed by further evaluation in a system such a human saphenous vein 'organ' culture, may prove more informative than animal studies, as these models of human VSMC migration and proliferation may more closely resemble the human *in vivo* VSMC environment [46].

5.2 VSMC phenotypic diversity

In association with disease, human VSMCs are heterogeneous. It is likely that different VSMC phenotypes represent populations of cells with specific functions. For example, in the atherosclerotic plaque, various VSMC sub-populations may exist which exhibit properties indicative of their involvement in: (1) vascular repair - the matrix producing, contractile-like cells of the fibrous cap; (2) calcification - the subset of VSMCs which express bone-associated proteins; (3) lipid accumulation - oil-red O positive VSMCs and (4) fibrous cap degeneration - apoptotic VSMCs and matrix-metalloproteinase producing VSMCs [20, 38, 47, 50, 51]. In order to identify specific markers of these different phenotypes and analyse their specific functional properties, human tissue culture models which mimic these various phenotypes *in vitro* are required. Although human culture systems are not as well defined as rodent models, some progress has been made. For example, in our laboratory, we have developed a model of vascular calcification where human VSMCs retract into multicellular nodules and deposit hydroxyapatite, the main type of calcium salt found in the vessel wall [21]. This model allows us to investigate the potential effects of various factors on the development of human vascular calcification. Additionally, in primary culture, cells derived from the atherosclerotic intima display a greater propensity to undergo apoptosis in low serum conditions than cells derived from the normal media [20]. These cells have therefore been used to study the mechanisms which regulate VSMC apoptosis using time-lapse video microscopy and molecular analysis.

The identification of stable human VSMC phenotypes *in vitro* will enable us to use differential cDNA screening systems to isolate specific markers to identify functional phenotypes *in vivo*. If different stable cultures could be established *in vitro*, then these cells could be used in studies to determine factors that regulate transitions between different functional phenotypes. This is of fundamental importance, as it is likely that local environmental factors in the vessel wall dictate the function of any given VSMC. Primary cell cultures are more likely to mimic cells *in vivo* than immortalized human VSMC lines which have been generated either by transfecting VSMC with plasmids containing SV40 [52] or by transforming cells with virus DNA [53]. These permanent lines retain expression of some smooth muscle-like traits, but they do not appear to resemble a functional *in vivo* phenotype [45]. Moreover, it has already been established that primary human VSMCs in culture are resilient enough to be used in a broad range of studies including co-culture systems where, for example, the effects of macrophages or endothelial cells on VSMC apoptosis, matrix production and other functions can be tested [54, 55]. Additionally, specialist culture systems for measuring effects of mechanical strain, flow and stretch on VSMCs have been successfully established.

5.3 VSMC contraction/differentiation

One area of research in which studies of human VSMCs have lagged behind animal studies is the identification of factors that influence and regulate VSMC differentiation. The isolation of embryonic and neonatal human VSMCs for tissue culture is rare and therefore these studies must still rely on the use of animal model systems. These models include murine cell lines which can be stimulated to differentiate towards a smooth muscle-like cell [56], and mesenchymal stem cells derived from human bone-marrow which also differentiate into cells which express smooth muscle specific markers [57]. The pinnacle of VSMC differentiation is its ability to contract, as observed in normal medial cells. To date, this phenotype has not been modeled in tissue culture as all cells modulate when they are removed from the environment of the vessel wall. Several reports describe attempts to reproduce the contractile phenotype of VSMCs in culture. These conditions include serum starvation or heparin treatment [58], coating culture flasks with laminin [59], plating cells on Matrigel [60] or in 3D collagen gel culture [19], while some recent progress has been made in mimicking differentiated cells in pig and rabbit models [39]. However, these are successful in maintaining only some aspects of the contractile phenotype. To date, only organ cultures of human vessels have been successful in measuring vasoconstriction and vasorelaxation in response to agonists [61], although one study has demonstrated contraction in a subset of human VSMCs in vitro [39a].

5.4 Human VSMCs and gene transfer

Gene transfer into blood vessels is a promising new approach to the treatment of vascular disease. Human VSMC cultures can be transfected with plasmid vectors, albeit at very low frequency, and infected at very high frequency with a variety of viral vectors. Therefore the potential exists to use modified VSMCs in cell transplant experiments, or to transfer genes using viral systems to modify the *in vivo* environment/phenotype of disease-associated VSMCs. The feasibility of the use of such systems in adult human blood vessels has recently been addressed by Rehkter et al. [62]. They incubated rings of human vessels with adenoviral vectors and found expression of the transgene in VSMCs as well as other cells in the normal and atherosclerotic vessels. These organ cultures are not an *in vivo* system with blood flow or arterial pressure, but provide a simplified culture system for the study of VSMC properties.

6. SUMMARY AND CONCLUDING REMARKS

The VSMCs in the normal artery wall have a contractile function *in vivo*. When VSMCs are removed from the media and grown in culture, contractile markers are reduced and the cells express a different repertoire of proteins. Cultured VSMCs therefore lose many of the characteristics of VSMCs in the normal vessel wall and resemble more closely those found in association with vascular disease. Moreover, cells *in vitro* and in association with disease are heterogeneous; they display different properties and different gene expression patterns, and therefore represent functionally distinct phenotypes. Progress has been made in establishing human VSMC culture systems and it is now possible to routinely culture VSMCs from many sources and from both normal and diseased tissue. The next challenge for the VSMC biologist will be to make human cells as versatile as those derived from rodents by establishing clear *in vitro* models of different functional phenotypes. It is hoped that future studies will reveal the factors that regulate VSMC differentiation and phenotypic transition so that we can gain insight into the regulation of VSMC plasticity. This ultimately will reveal potential factors for targeting in interventional treatments for vascular diseases.

REFERENCES

1. Chamley-Campbell JH, Campbell GR and Ross R. The smooth muscle cell in culture. Physiol Rev 1979, 59:1-61.
2. Ross R. The pathogenesis of atherosclerosis - an update. N Eng J Med 1986, 314(8):488-500.
3. Mosse PRL, Campbell GR, Wang ZL, and JH Campbell. Smooth muscle phenotypic expression in human carotid arteries I. Comparison of cells from diffuse intimal thickenings adjacent to atheromatous plaques with those of the media. Lab Invest 1985; 53: 556-562.
4. Campbell GR, and Campbell JH. Smooth muscle cell phenotypic changes in arterial wall homeostasis: implications for the pathogenesis of atherosclerosis. Exp Mol Pathol 1985, 42;139-162.
5. Schwartz SM, DeBois D, and O'Brien ERM. The intima: soil for atherosclerosis and restenosis. Circ Res 1995, 77:445-465.
6. Schwartz SM, Campbell GR, and Campbell JH. Replication of smooth muscle cells in vascular disease. Circ Res 1986, 58:427-444.
7. Jonasson L, Holm J, Skalli O, Gabbiani G, and Hansson GK. Expression of Class II transplantation antigen on vascular smooth muscle cells in human atherosclerosis. J Clin Invest 1985, 76:125-131.

8. Hansson GK, Jonasson L, Holm J, and Claesson-Welsh L. Class II MHC antigen expression in the atherosclerotic plaque: smooth muscle cells express HLA-DR, HLA-DQ and the invariant gamma chain. Clin Exp Immunol 1986, 64:261-268.
9. Ross R, and Glomset J. The pathogenesis of atherosclerosis. Part 1. N Engl J Med 1976, 295:369-377.
10. Ross R, and Glomset J. The pathogenesis of atherosclerosis. Part 2. N Engl J Med 1976, 295:420-428.
11. Davies MJ. Acute coronary thrombosis - The role of plaque disruption and its initiation and prevention. Eur Heart Journal, 1995, 16:3-7.
12. Frid MG, Dempsey EC, Durmowicz AG, and Stenmark KR. Smooth muscle cell heterogeneity in the pulmonary and systemic vessels: Importance in vascular disease. Arterioscler Thromb Vasc Biol 1997; 17:1203-1209.
13. Bochaton-Piallat M-L, Ropraz P, Gabbiani F, and Gabbiani G. Phenotypic heterogeneity of rat arterial smooth muscle cell clones: implications for the development of experimental intimal thickening. Arterioscler Thromb Vasc Biol 1996, 16:815-820.
14. Neuville P, Geinoz A, Benzonana G, Redard M, Gabbiani F, Ropraz P, and Gabbiani G. Cellular retinol-binding protein-1 is expressed by distinct subsets of rat arterial smooth muscle cells in vitro and in vivo. Am J Pathol 1997, 150:509-521.
15. Babaev VR, Bobryshev YV, Stenina OV, Tararak EM, and Gabbiani G. Heterogeneity of smooth muscle cells in atheromatous plaque of human aorta. Am J Path 1990, 136:1031-1042.
16. Glukhova MA, Kabakov AE, Frid MG, Ornatsky OI, Belkin AM, Mukhin DN, Orekhov AN, Koteliansky VE, and Smirnov VN. Modulation of human aorta smooth muscle cell phenotype: A study of muscle-specific variants of vinculin, caldesmon, and actin expression. Proc Natl Acad Sci USA 1988, 85:9542-9546.
17. Dartsch PC, Weiss HD, and Betz E. Human vascular smooth muscle cells in culture: growth characteristics and protein pattern by use of serum-free media supplements. Eur J Cell Biol 1990, 51:285-294.
18. Fujita H, Shimokado K, Yutani C, Takaichi S, Masuda J, and Ogata J. Human neonatal and adult vascular smooth muscle cells in culture. Exp Mol Path 1993, 58:25-39.
19. Scherberich A, Moog S, HaanArchipoff G, Azorsa DO, Lanza F, and Beretz A. Tetraspandin CD9 is associated with very late-acting integrins in human vascular smooth muscle cells and modulates collagen matrix reorganisation. Arterioscler Thromb Vasc Biol 1998, 18:1691-1697.
20. Bennett MR, Littlewood TD, Schwartz SM, and Weissberg PL. Increased sensitivity of human vascular smooth muscle cells from atherosclerotic plaques to p53-mediated apoptosis. Circ Res 1997, 81:591-599.
21. Proudfoot D, Skepper JN, Shanahan CM, and Weissberg PL. Calcification of human vascular cells in vitro is correlated with high levels of matrix Gla protein and low levels of osteopontin expression. 1998; 18:379-388.
22. Shanahan CM, Weissberg PL, and Metcalf JC. Isolation of gene markers of differentiated and proliferating vascular smooth muscle cells. Circ Res 1993; 73:193-204.
23. Newman CM, Bruun BC, Porter KE, Mistry PK, Shanahan CM, and PL Weissberg. Osteopontin is not a marker for proliferating human vascular smooth muscle cells. Arterioscler Thromb Vasc Biol 1995; 15: 2010-2018.
24. Shioi A, Nishizawa Y, Jono S, Koyama H, Hosoi M, and H Morii. β-glycerophosphate accelerates calcification in cultured bovine vascular cells. Arterioscler Thromb Vasc Biol 1995; 15: 2003-2009.

25. Rutstein DD, Ingenito EF, Craig JM, and Martinelli M. Effects of linolenic and stearic acids on cholesterol-induced lipoid deposition in human aortic cells in tissue culture. The Lancet 1958, I, 545-552.
26. Gittenberger-de Groot AC, DeRuiter MC, Bergwerff M, and Poelmann RE. Smooth muscle cell origin and its relation to heterogeneity in development and disease. Arterioscler Thromb Vasc Biol 1999, 19:1589-1594.
27. Topouzis S, and Majesky MW. Smooth muscle lineage diversity in the chick embryo - Two types of aortic smooth muscle cell differ in growth and receptor-mediated transcriptional responses to transforming growth factor-beta. Developmental Biol, 1996, 178:430-445.
28. Schlingemann RO, Rietveld FJR, W.deWaal RM, Ferrone S, and Ruiter DJ. Expression of the high molecular weight melanoma-associated antigen by pericytes during angiogenesis in tumours and in healing wounds. Am J Path 1990, 136:1393-1405.
29. Lee CS, Patton, WF, Chung-Welch N, Chiang ET, Spofford KH, and Shepro D. Selective propagation of retinal pericytes in mixed microvascular cell cultures using L-leucine-methyl ester. BioTechniques 1998, 25:482-494.
30. Dartsch PC, Voisard R, Bauriedel G, Hofling B, and Betz E. Growth characteristics and cytoskeletal organisation of cultured smooth muscle cells from human primary stenosing and restenosing lesions. Arteriosclerosis 1990, 10:62-75.
31. Torzewski J, Oldroyd R, Lachmann P, Fitzsimmons C, Proudfoot D, and Bowyer DE. Complement-induced release of monocyte chemotactic protein-1 from human smooth muscle cells. Arterioscler Thromb Vasc Biol 1996, 16:673-677.
32. Cheng GC, Briggs WH, Gerson DS, Libby P, Grodzinsky AJ, Gray ML, and Lee RT. Mechanical strain tightly controls fibroblast growth factor-2 release from cultured human vascular smooth muscle cells. Circ Res 1997, 80:28-36.
33. Bennett MR, MacDonald K, Chan SW, Boyle JJ, and Weissberg PL. Cooperative interactions between RB and p53 regulate cell proliferation, cell senescence, and apoptosis in human vascular smooth muscle cells from atherosclerotic plaques. Circ Res 1998, 82:704-712.
34. Thyberg J, Hedin U, Sjolund M, Palmberg L, and Bottger BA. Regulation of differentiated properties and proliferation of arterial smooth muscle cells. Arteriosclerosis 1990, 10:966-989.
35. Björkerud S, Björkerud B, and Joelsson M. Structural organisation of reconstituted human arterial smooth muscle tissue. Arterioscler. Thromb 1994; 14: 664-651.
36. Gimbrone MA, and Cotran RS. Human vascular smooth muscle in culture: growth and ultrastructure. Lab Invest 1975; 33: 16-27.
37. Schor AM, Allen TD, Canfield AE, Sloan P, and Schor SL. Pericytes derived from the retinal microvasculature undergo calcification in vitro. J Cell Science 1990; 97: 449-461.
38. Shanahan CM, Cary NRB, Salisbury JR, Proudfoot D, Weissberg PL, and Edmonds ME. Medial localisation of mineralization-regulating proteins in association with Mönckeberg's sclerosis: Evidence for smooth muscle cell-mediated vascular calcification. Circulation 1999;100:2168-2176.
39. Shirinsky VP, Birukov KG, Sobolevsky AV, Vedernikov YP, Posin EY, and Popov EG. Contractile rabbit aortic smooth muscle cells in culture – preparation and characterization. Am J Hypertension 1992; 5 (no.6, pt2 SS): S124-S130.
39a. Li SH, Sims S, Jiao Y, Chow LH, and Pickering JG. Evidence from a novel human cell clone that adult vascular smooth muscle cells can convert reversibly bewteen noncontractile and contractile phenotypes. Circ Res 1999;85:338-348.

40. Skalli O, Ropraz P, Trzeciak A, Benzonana G, Gillesen D, and Gabbiani G. A monoclonal antibody against α-smooth muscle actin: a new probe for smooth muscle differentiation. J Cell Biol 1986, 103:2787.
41. Arciniegas E, Sutton AB, Allen TD, and Schor AM. Transforming growth factor beta-1 promotes the differentiation of endothelial cells into smooth muscle-like cells in vitro. J Cell Sci, 1992, 103:521-529.
42. Shi Y, O'Brien JE, Frad A, and Zalewski A. Transforming growth factor beta-1 expression and myofibroblast formation during arterial repair. Arterioscler Thromb Vasc Biol 1996, 16:1298-1305.
43. Nayak RC, Berman AB, George KL, Eisenbarth GS, and King GL. A monoclonal antibody (3G5)-defined ganglioside antigen is expressed on the cell surface of microvascular pericytes. J Exp Med 1988, 167:1003-1015.
44. Shepro D, and Morel NML. Pericyte physiology. FASEB J 1993, 7:1031-1038.
45. Bonin LR, Madden K, Shera K, Ihle J, Matthews C, Aziz S, Perez-Reyes N, McDougall JK, and Conroy SC. Generation and characterisation of human smooth muscle cell lines derived from atherosclerotic plaque. Arterioscler Thromb Vasc Biol 1999, 19:575-587.
46. Kranzhofer A, Baker AH, George SJ, and Newby AC. Expression of tissue inhibitor of metalloproteinase-1, -2, and -3 during neointima formation in organ cultures of human saphenous vein. Arterioscler Thromb Vasc Biol 1999, 19:255-265.
47. Sukhova GK, Shi GP, Simon DI, Chapman HA, and Libby P. Expression of the elastolytic cathepsins S and K in human atheroma and regulation of their production in smooth muscle cells. J Clin Invest 1998, 102:576-583.
48. Proudfoot D, Fitzsimmons C, Torzewski J, and Bowyer, DE. Inhibition of human arterial smooth muscle cell growth by human monocyte/macrophages: a co-culture study. Atherosclerosis 1999, 145:157-165.
49. Thyberg J. Differentiated properties and proliferation of arterial smooth muscle cells in culture. Int Rev Cytol 1996; 169:183-265
50. Shanahan CM, and Weissberg PL. Smooth muscle cell heterogeneity: Patterns of gene expression in vascular smooth muscle cells in vitro and in vivo. Arterioscler Throm Vasc Biol 1998, 18:333-338.
51. Stary HC, Blankenhorn DH, Chandler AB, Glagov S, Insull W, Richardson M, Rosenfeld ME, Schatter SA, Schwartz CJ, Wagner WD, and Wissler RW. A definition of the intima of human arteries and of its atherosclerosis-prone regions. Arterioscler Thromb 1992, 12:120-133.
52. Legrand A, Greenspan P, Nagpal ML, Nachtigal SA, and Nachtigal M. Characterisation of human vascular smooth muscle cells transformed by the early genetic region of SV40 virus. Am J Pathol 1994;139:629-640.
53. Perez-Reyes N, Halbert CL, Smith PP, Benditt EP, and McDougal JK. Immortalisation of primary human smooth muscle cells. Proc Natl Acad Sci USA 1992;89:1224-1228.
54. Fitzsimmons C, Proudfoot D, and Bowyer DE. Monocyte prostaglandins inhibit procollagen secretion by human vascular smooth muscle cells:implications for plaque stability. Atherosclerosis 1999, 142:287-293.
55. Boyle JJ, Bowyer DE, Proudfoot D, Weissberg PL, and Bennett MR. Human monocyte/macrophages induce human vascular smooth muscle cell apoptosis in culture. Circulation, 1998, 98, No.17 SS:3142.
56. Suzuki T, Kim HS, Kurabayashi M, Hamada H, Fujii H, Aikawa M, Watanabe M, Sakomura Y, Yazaki Y, and Nagai R. Preferential differentiation of P19 mouse embryonal carcinoma cells into smooth muscle cells. Circ Res 1996, 78:395-403.

57. Li J, Sensebe L, Herve P, and Charbord P. Nontransformed colony-derived srtomal cell lines from normal human marrows. II. Phenotypic characterisation and differentiation pathway. Exp Haematol 1995, 23:133-141.
58. Desmouliere A, Rubbia-Brandt L, and Gabbiani G. Modulation of actin isoform expression in cultured arterial smooth muscle cells by heparin and culture conditions. Arterioscler Thromb 1991, 11:244-253.
59. Hedin U, Bottger BA, Forsberg E, Johnasson S, and Thyberg J. Diverse effects of fibronectin and laminin on phenotypic properties of cultured arterial smooth muscle cells. J Cell Biol 1988, 107:307-319.
60. Li X, Tsai P, Wieder ED, Kribben A, Van Putten V. Schrier RW, and Nemenoff RA. Vascular smooth muscle cells grown on Matrigel: a model of the contractile phenotype with decreased activation of mitogen-activated protein kinase. J Biol Chem 1994, 269:19653-19658.
61. Macguire JJ, and Davenport AP. Endothelin receptor expression and pharmacology in human saphenous vein graft. British J Pharmacol 1999, 126:443-450.
62. Rehkter MD, Simari RD, Work CW, Nabel GJ, Nabel EG, and Gordon D. Gene transfer into normal and atherosclerotic human blood vessels. Circ Res 1998, 82:1243-1252.

Chapter 5

Skeletal Muscle

Peter FM van der Ven
Department of Cell Biology, University of Potsdam, D14471 Potsdam, Germany. Tel: 0049-331- 977- 4856; Fax: 0049-331- 977- 4861; E-mail: pvdven@rz.uni-potsdam.de

1. INTRODUCTION

Cellular motility and cell movement in organisms as unrelated as amoebae, invertebrates and mammals involve the interaction of myosin and actin filaments. The process of muscle contraction is among the most intensely studied functions dependent on this interaction. In the human body, three distinct types of muscular tissues can be discerned. Smooth muscles are an integral part of the vascular system and are found, for instance, within the digestive system, the bladder and the uterus. The striated myocytes of the heart control blood circulation, while the second class of striated muscles enables us to move (skeletal muscles), and help us speak (tongue) and breathe (diaphragm). Of all the different tissues and organs in the human body, skeletal muscles are by far the most frequent organ: more than 600 different muscles comprise up to 50% of total body weight. A unique characteristic of skeletal muscle is that it is not made up of individual cells, but consists of large multinucleate syncytia, the myofibers, each of which can be several centimeters long. During embryogenesis, mononuclear precursor cells, or myoblasts, fuse to form these huge muscle fibers. This process is mimicked in later life during regenerative processes subsequent to muscle damage, whereby quiescent mononuclear stem cells (so-called satellite cells) located between the sarcolemma and the basal lamina of the muscle fiber are activated. These cells begin to proliferate and either fuse with each other into novel myotubes, or they fuse with damaged muscle fibers. Both myoblasts and satellite cells can be isolated from the body and grown in tissue culture, and in optimal culture media they will fuse and differentiate into mature, spontaneously-contracting myotubes. Cultures derived from normal and diseased human skeletal muscle cells therefore

M.R. Koller, B.O. Palsson and J.R.W. Masters (eds.), Human Cell Culture Vol V, 65–101.

provide an excellent model to study several aspects of early muscle development under normal and pathological conditions.

1.1 Control of Muscle Development in Vertebrates

The precursor cells of the epaxial skeletal muscles (body and trunk muscles) arise from the somites, repetitive epithelial structures that during gastrulation are transiently formed immediately adjacent to the neural tube from the paraxial mesoderm [1, 2]. The ventromedial region of the somites further differentiates into the sclerotome, whereas the dorsolateral domain forms the dermomyotome, that, most likely in a multistep process [2], gives rise to a second layer, the myotome. Within specific portions of the dermomyotome and the myotome, muscle precursor cells develop which during later stages of embryogenesis produce skeletal muscle in well defined regions of the body. Cells from the medial part of the dermomyotome migrate to the medial part of the myotome and differentiate into epaxial muscle that will yield back muscles. Similarly, cells from the lateral dermomyotome translocate to the lateral myotome and give rise to hypaxial muscle that will produce the ventral body wall muscles from the thorax and the abdomen. The hypaxial muscles from the limbs, however, are also derived from precursor cells from the dermomyotome, but in contrast to the body wall musculature, these cells do not use the myotomes as an intermediate location. These cells are often translocated over considerable distances through the body before they proliferate further and differentiate into the muscle of the limbs, the diaphragm, and typically, the tip of the tongue. The rest of the tongue and the other muscles of the head are not only derived from migratory cells from (rostral) somites, but also from paraxial head mesoderm and prechordal mesoderm.

1.2 Regulation of Muscle Cell Determination and Migration

Our understanding of skeletal muscle lineage determination has enormously increased since the identification of the myogenic determination factor genes. The four members of this MyoD family, or basic helix-loop-helix (bHLH) transcription factor family, MyoD [3], Myf-5 [4], myogenin [5, 6] and MRF4 [7], also called myf-6 [8] or herculin [9], all have specific functions in muscle determination and differentiation (Figure 1) that became apparent after the introduction of null mutations in mice [10-16], as reviewed in [17-19]. MyoD and Myf-5 are responsible for commitment of mesenchymal precursor cells to the myogenic lineage. More specifically, MyoD is essential for migratory muscle cell lineages, whereas Myf-5 is necessary for determination of lineages that are derived from the myotomes [20, 21]. In mice lacking both genes, no muscle precursor cells are generated. In contrast, myogenin and MRF4 are not required for skeletal muscle lineage determination, but are active downstream of the other two members of the MyoD family: myogenin is responsible for fusion of muscle cells and the formation of myotubes, whereas MRF4 plays a role in the subsequent differentiation of myotubes into mature myofibers [12, 13, 15]. In adult muscle fibers, MRF4 is the predominant transcript [7]. Recent reviews on this topic, the regulation of expression of bHLH transcription factors and the role of integrins and cadherins in muscle development and migration, include references 22-25, all collected in a special issue of Cell and Tissue Research.

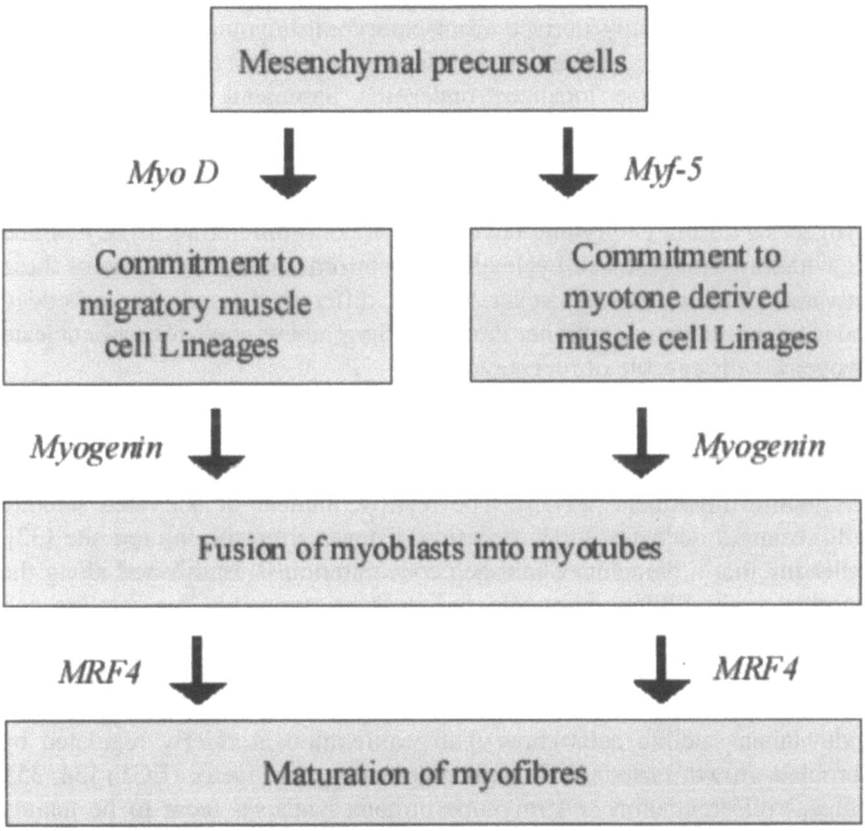

Figure 1. Functions of myogenic determination factors in muscle determination and differentiation. For more details see text.

1.3 Regeneration of Skeletal Muscle

Although it has long been known that myotube nuclei are not capable of mitosis following the fusion of mononucleate myoblasts into myofibers during late gestation and neonatal development, these myofibers grow and the number of nuclei per fiber increases in the postnatal period. Furthermore, nuclei from damaged muscle fibers can be replaced during regenerative processes, and upon overuse of muscle, not only the size of muscle fibers increases, but also the number of nuclei per fiber. The origin of these nuclei was unclear until the identification of satellite cells [26, 27], and the suggestion that the nuclei of these cells might replace the nuclei of damaged muscle fibers during regenerative processes [26]. This suggestion was confirmed, and we now know that these satellite cells are mitotically quiescent cells, initially derived from embryonic myoblasts that withdrew from the cell cycle, adhered to developing myotubes and during further differentiation became localized under the basement membrane of the myofiber. The cells are activated and start to proliferate when existing muscle fibers are injured and need to be repaired or replaced, or existing myofibers need to grow [28-30]. Muscle regeneration then recapitulates myogenesis during embryonic development, i.e. proliferating myocytes line up, withdraw from the cell cycle and fuse to form myotubes. Some of these activated satellite cells, however, do not differentiate and are embedded under the basement membrane, thus providing a new pool of mononucleate myogenic cells capable of regeneration.

The mechanisms by which quiescent satellite cells are activated *in vivo* are not well understood. Proliferating cells are not only recruited from sites close to the injury, but also from sites distant to the trauma or even from neighboring myofibers [31-33]. The relative number of activated satellite cells seems to decrease with increasing distance from the injured site [32], indicating that a gradient of mitogen concentration is established along the muscle length. Which mitogenic factors are responsible for satellite cell activation on the one hand, and satellite cell differentiation on the other hand, is difficult to examine *in vivo*. Therefore, most studies on these topics were performed with cultured satellite cells (see below). The few studies with human satellite cells show that proliferation is chiefly regulated by fibroblast growth factor (FGF) and epidermal growth factor (EGF) [34, 35], whereas differentiation and myotube protein synthesis seem to be mainly stimulated by insulin and insulin-like growth factors (IGFs) [34, 36, 37]. Evidence for the *in vivo* and *in vitro* differentiation-stimulating effect of IGF-I was recently provided in several studies [38-41], and the mechanism of the stimulatory effect of IGF-I, which was mimicked by insulin in combination with dexamethasone, was shown to be mediated by a calcineurin signaling pathway [42, 43].

2. TISSUE PROCUREMENT AND PROCESSING

2.1 Tissue Procurement

As sources for the isolation of myoblasts or satellite cells, theoretically all skeletal muscles are suitable, although the cell yield and the replicative capacity of the isolated cells may vary considerably with the site of muscle biopsy and the age of the donor [44, 45] (see below). Fresh non-fixed and non-frozen muscle specimens are usually obtained during corrective surgery or from biopsies taken for diagnostic purposes. In some scientific reports, the authors succeeded in finding volunteers willing to donate a small piece of muscle tissue [46]. The risks associated with a muscle biopsy are usually minimal, although the donor might experience considerable postoperative discomfort. Biopsies are usually taken under local anesthesia using scissors or a muscle biopsy clamp, or by percutaneous needle biopsy. Although it is no problem to isolate satellite cells from small needle biopsies, the total number of cells may be insufficient for the subsequently planned experiments. Alternatively, muscle specimens might be obtained during autopsy. Informed consent must be obtained from the donor or relatives, according to the local legislation.

Normal human muscle biopsies can usually be obtained relatively easily from a hospital with an orthopedic department. However, a major problem with the study of cultures of diseased skeletal muscle cells is that at the time of biopsy the diagnosis is often unknown, and the tissue is fixed or frozen for histopathological staining, electron microscopy or biochemistry. Therefore, collaborations with neurologists or orthopedics who are interested in biomedical research, and willing to cooperate in preparing a portion of the muscle biopsy for tissue culture, are important.

A highly praiseworthy initiative of the Friedrich-Baur-Institute of the University of Munich, Germany that is supported by the *Deutsche Gesellschaft für Muskelkrankheiten* (i.e. German Society for Muscle Diseases) is the establishment of a collection of primary human muscle cell cultures derived from normal and diseased human skeletal muscle biopsies. Researchers can support the work of the *Muskelbank* by sending biopsies for the isolation of satellite cells, that, in small quantities, are made available for biomedical research at universities and non-profit institutions. The establishment of a complete collection would be an enormous improvement to the resources available to study normal and diseased human skeletal muscle cells in culture. Normal human skeletal muscle cells can also be purchased from companies such as PromoCell (Heidelberg, Germany) and BioWhittaker (Walkersville, MD).

2.2 Tissue Processing

Probably due to the localization of the satellite cells underneath the basement membrane of the muscle fiber, muscle specimens can be stored for a conveniently long period. If the tissue is kept on ice in an appropriate buffer, such as sterile Dulbecco's PBS (DPBS; 145 mM NaCl, 5.4 mM KCl, 5 mM Na_2HPO_4, 25 mM glucose, 25 mM sucrose, 50 IU/ml penicillin and 10 µg/ml streptomycin) [47], or Ham's F12 medium supplemented with 10mM HEPES/NaOH (pH 7.2) [44], the tissue specimens can be stored before processing for at least three days. Slightly fewer viable cells will be isolated from the biopsies (Table 1), but storage has no effect on proliferation and differentiation rates. After three days, however, the cell yield seems to drop considerably [44]. This allows transportation over long distances, and considering the time-consuming procedure for the enzymatic isolation of satellite cells described below, better planning of the experiments.

Table 1. The effect of storage of muscle biopsies in Ham's F12 medium supplemented with 10 mM HEPES, pH 7.2 at 4°C on the yield of satellite cells upon enzymatic dissociation [44].

Duration of storage	Relative cell yield (%)
<12 hours	100
1-3 days	66.8
>4 days	40.1

As an alternative, freshly biopsied muscle tissue may be frozen in liquid nitrogen. Approximately 3 mm^3 pieces of muscle tissue are slowly frozen in sterile freezing vials in freezing medium consisting of 10% DMSO as a cryoprotectant in culture medium or FCS [48, 49 and our unpublished results] and our unpublished results. For freezing of cells, the temperature should drop approximately 1°C per minute. One can also place the vials at –20°C for one hour and subsequently overnight at –70°C before storing them in liquid nitrogen (our unpublished results). Many laboratories put their samples in a homemade container made of insulated cardboard, plastic or polystyrene foam and place it directly inside a –70°C freezer. This method is, however, not very reliable, and might cause considerable cell death. A relatively cheap and better method for freezing tissue fragments and cells uses a simple freezing container obtainable from Nalgene (Naperville, IL), filled with isopropyl alcohol. Cryovials can be placed in a vial holder in such a way that the alcohol functions as an insulator and the complete device can be placed at –70 °C.

For the isolation of myoblasts or satellite cells, one or more enzymes are usually used to dissociate the tissue. Numerous different protocols using a single enzyme or combinations of enzymes, such as trypsin [50], collagenase and trypsin [47, 51, 52], collagenase and pronase [44, 53], or pretreatment with collagenase and hyaluronidase, and subsequent further digestion with trypsin [54], have been described. A modification of a protocol initially developed by Yasin et al. [47] and applied successfully over many years in our laboratory, is described below.

Fresh muscle biopsies are submerged in cold, sterile DPBS supplemented with 100 U/ml penicillin and 10 μg/ml streptomycin immediately subsequent to surgical removal from the body. Fat and connective tissue are removed with sterile forceps, scissors or dissecting knives. Approximately 250mg of the remaining tissue is rinsed a few times in DPBS, cut into small fragments and added to 5 ml DPBS containing 0.15% trypsin (1:250, Life Technologies), 0.1% collagenase (tested batch, usually Sigma C-2139 works well), and 0.1% BSA (BDH Laboratory Supplies). The tissue is incubated in a slowly shaking water bath at 37 °C. After 15 minutes, the buffer is carefully removed and discarded, leaving residual tissue fragments in the container. At this stage of the procedure, it is mostly erythrocytes that are removed from the tissue fragments. After the addition of fresh enzyme solution (5 ml), the tissue fragments are processed further with sterile needles and the container is placed again in the water bath for 15 minutes. Dissociated cells are collected in an equal volume of 10% FCS in DMEM supplemented with antibiotics as above. This dissociation step is repeated two more times. Subsequently, 5 ml of DMEM containing FCS is added to the remaining tissue fragments, and these are gently forced ten times in and out of a 10 ml plastic tissue culture pipette. Supernatants with dissociated satellite cells are collected in medium, leaving larger tissue fragments at the bottom of the container. Cell suspensions are pooled and filtered through 300 μm and 36 μm nylon mesh filters to remove residual tissue fragments. The cells are collected by centrifugation, counted and plated into tissue culture dishes at a density of approximately 50,000 cells per 35 mm culture dish in culture medium containing a high concentration of serum. We currently use DMEM (with 4.5 mg/ml glucose) supplemented with 20% selected FCS, 2% Ultroser G, 2 mM glutamine, 100 U/ml penicillin and 10 μg/ml streptomycin. The next day, cellular debris and non-adherent cells are removed by washing with DPBS and fresh medium is added to the cells. Every two to three days, the culture medium is replaced with fresh medium, and the cells should be passaged before reaching confluence.

An alternative, also widely used method for the cultivation of human skeletal muscle cells is the so-called explant-reexplantation technique initially developed in the laboratory of Askanas and Engel [48, 55]. The establishment of muscle cultures is much less time-consuming, since muscle tissue is not treated with enzymes, but simply cut into small, 1mm^3 pieces that are placed into Petri dishes coated with a mixture of Factor VIII-depleted human plasma and gelatin, allowing cells to grow out from the tissue fragment. After two weeks, the explants are reexplanted into fresh, coated petri dishes with culture medium. While in the original protocol three different media, one of them containing human serum, were used [48], in more recent publications a single medium consisting of F14 medium [56] with 6 mg/ml glucose supplemented with 10% FCS, 25 ng/ml FGF, 10 ng/ml EGF, and 10 µg/ml insulin was used [35, 57, 58]. Since the cells that grow out of the primary explants are mainly fibroblasts, these cells have to be left in the original petri dish. Outgrowing cells from the re-explanted tissue fragment are mainly myoblasts that are allowed to proliferate and differentiate without changing the culture medium to a low-nutrition medium. A combination of the explant technique and mass monolayer culture includes trypsinization of the cells that grow out of the explants in order to prepare secondary cultures on a gelatin coated substrate [59].

A further method for studying proliferation and fusion of satellite cells worth mentioning here, although until now not applied on human muscle fibers, is culture of isolated single fibers. This culture system was developed by Bischoff [60], who used a modification of a protocol described by Bekoff and Betz [61]. It allows the examination of satellite cells that remain attached to the muscle fiber, and therefore can be considered to better resemble the *in vivo* situation. In contrast to the monolayer mass cultures that start with the liberation of satellite cells from muscle tissue, in this method muscle tissue is only mildly enzymatically digested, leaving the muscle fibers intact with the satellite cells still lying between the basal lamina and the sarcolemma. It proved to be important to remove colostripain from crude collagenase, since colostripain damages the basal lamina [60]. The obvious advantage is the contact of the satellite cells with the muscle fiber sarcolemma and the basal lamina that cannot be reproduced in monolayer cultures. The method was used to study feedback both between the sarcolemma and satellite cells, and between the basement membrane and satellite cells [62, 63]. Furthermore, it was shown to be highly suitable for studies concerning the response of satellite cells to the addition of mitogens, growth factors and hormones [62-65]. A drawback of the method is that only small, intact muscle fibers can be studied, since in damaged muscle fibers regenerative processes will be initiated. Rosenblatt and co-workers adapted the protocol to allow treatment of larger muscle fibers and showed that the few satellite cells derived from a single myofiber suffice for most experiments. With this method, specific populations of satellite cells that are associated with individual slow or fast muscle fibers can be studied [66, 67]. Again, this culture system was not optimized for human myofibers, and the technical difficulties associated with the fragility of single human myofibers might make the method unsuitable [68].

One can also take advantage of the regenerative processes that are initiated after muscle fiber damage and study regeneration in explants of human muscle [69, 70]. Muscle samples, either fresh or frozen in 10% DMSO, 20% human placental cord serum in Hank's balanced salt solution, are cut into 0.7 - 1.0 mm long pieces containing a bundle of up to 20 muscle fibers. The tissue fragment is placed in culture medium consisting of 65% DMEM, 25% human placental cord serum, 10% chick embryo extract and 6 mg/ml glucose. The satellite cells of the damaged fibers are activated, start to regenerate and fuse to form new myotubes along and within the basement membrane of the old fibers that degenerate. The complete process of regeneration can be studied by immunocytochemistry and electron microscopy. Small fibers that are not damaged during tissue preparation, or that are resealed directly after preparation may, however, persist in the cultures [71].

2.3 Donor-to-Donor and Muscle-to-Muscle Variability

Depending on the planned experiments, different numbers of cells are required. In cases where large numbers of cells are needed, the choice of donor and the site of biopsy might be important. Although it was reported that a single satellite cell can be expanded to at least 10 million cells [50], which would enable extensive biochemical studies with cells isolated from a small biopsy, the replicative capacity varies considerably with donor age. Under identical *in vitro* conditions, the average number of progeny produced was inversely proportional to donor age. Hence, the proliferation potential of satellite cells decreases with age [72, 73]. While the average replication capacity was reported to be 45 doublings [74], this capacity may vary from less than 10 doublings for satellite cells isolated from a muscle biopsy from a 62 year old person to 65 doublings for cells from a newborn child [72]. Furthermore, the yield of viable satellite cells per mg of tissue decreases with increasing donor age [44]. The sex of the donor did not have any influence on the cell yield or proliferation and fusion rates. Since all satellite cells seem to have the potential to proliferate and fuse [44, 75], intrinsically they are all appropriate for most studies. The limited capacity of cells isolated from biopsies from older donors might, however, not make them the cells of choice for experiments requiring very large numbers of cells.

In choosing the biopsy site, one should consider that oxidative muscle fibers have a much higher density of satellite cells than do glycolytic muscle fibers [76], and therefore, the cell yield per gram of biopsy may vary considerably depending on the biopsied muscle [44]. The question whether satellite cells derived from individual muscles are a uniform population, or if the cells show a diversity in their characteristics, is yet to be answered. The latter would mean that cultures derived from different muscles may vary. Although several groups reported that avian [77, 78], rabbit [79] and human [80] satellite cells are pre-programmed to form myotubes of a certain type, other authors found no evidence for the existence of different fast and slow satellite cell lineages in rat [81] and human postnatal skeletal muscle [82, 83]. This discrepancy possibly can be explained by differences in culture conditions, indicating that extrinsic factors, like the choice of the substrate, may play a role in determining myotube diversity [84].

3. CULTURE TECHNIQUES

3.1 Choice of Culture Medium

An optimized culture medium is extremely important to achieve optimum proliferation and differentiation of human muscle cells. In the early 1970s, a variety of techniques were developed that subsequently were adapted to find the most favorable conditions for the experiments performed in each laboratory. In 1976, a list of 15 different culture media applied in different laboratories was presented [59]. Many culture media are suitable for culturing human skeletal muscle cells. Firstly, one should decide whether to use a single medium for proliferation and differentiation of the cells, or separately use a high nutrition medium for proliferation and a low-nutrition medium for differentiation. In our laboratory, we use two separate media [85-87]. The cultures reach a higher density in a typical growth medium, differentiation of the myoblasts is more synchronous, and the decreased amount of nutrients in the medium seems to stimulate withdrawal from the cell cycle and accelerate differentiation [34, 88]. High nutrition media traditionally contain 15 or 20% FCS, often enriched with 0.5 or 2% chicken embryo extract, in synthetic culture media such as DMEM [47, 50, 75, 89], MEM [54, 90], MCDB 120 [34], Ham's F10 [91-93], Ham's F12 [44], medium 199 [94, 95] or mixtures of these media [96]. Further supplements have been reported to enhance human muscle proliferation and differentiation (e.g. rat brain extract) [51, 97]. The stimulating effects of sera, embryo extracts and brain extracts were found empirically, and are still not completely understood. In the search for improved media, it was found that sera and embryo extracts can, at least in part, be substituted by FGF,

EGF, insulin, dexamethasone, albumin and fetuin [34], or transferrin, insulin, FGF and EGF [35, 88]. The composition of a medium without sera or embryo extracts would be defined and thus would exclude variabilities due to the usage of different serum or extract batches. A defined medium was developed to allow serum-free growth of human skeletal muscle cells [34]. This medium consists of MCDB 120 nutrient mix, a medium in which the concentrations of all nutrients were optimized for human muscle satellite cells, supplemented with FGF, EGF, insulin, dexamethasone, BSA and fetuin. Addition of FCS to the culture medium, however, further improved growth, indicating that FCS contains yet unknown proliferation stimulating factors [34]. In this medium, cells hardly differentiate and they have to be transferred to a differentiation medium such as DMEM supplemented with 10 μg/ml insulin to allow fusion. Although fusion will occur in most low-nutrition media, such as 2% or 10% HS in DMEM, with or without CEE, other media were described to enhance differentiation. These include DMEM supplemented with 0.4% Ultroser G [51, 52], with 10 μg/ml insulin and 100 μg/ml human transferrin [88], or with 10 μg/ml insulin, 0.5 mg/ml BSA and 10 ng/ml EGF [93, 98]. The reported stimulatory effect of Ultroser G on the proliferation and, in low concentrations, also on the differentiation of human skeletal muscle cells [51, 52, 97], can probably be explained by the fact that most of the above mentioned stimulatory components are present in this medium substitute [51]. Recent reviews summarizing the multiple studies on the effects of growth factors and hormones on *in vitro* mammalian myoblast proliferation and differentiation include references [36] and [37] (see also Table 2).

3.2 Coating of Growth Surfaces

Primary human skeletal muscle cells grow and differentiate very well on uncoated plastic tissue culture surfaces [51, 97, 99, 100], and even on untreated glass [52, 85]. However, most researchers prefer treating the surface on which muscle cells are cultured either with gelatin [59, 80, 95], a mixture of human plasma and gelatin [48] or collagen [35, 47, 50, 54, 101]. Other researchers prefer Matrigel (Becton Dickinson), a mixture of several basement membrane components such as laminin, type IV collagen, heparan sulfate proteoglycan and entactin [80].

Table 2. Effects of growth factors and hormones on proliferation and differentiation of myoblasts and satellite cells. The data are based on several reports (for references see refs 29, 36 and 37). Therefore, some contradictory data appear in the table. 0: no effect; -: inhibitory effect; +: stimulatory effect; ?: no published data.

Factor	Proliferation	Differentiation
Insulin	0, +	+
Insulin-like growth factors (IGFs)	+	+
Growth hormone	0	0, +
Fibroblast growth factor-1, -2 (FGF-1, -2)	0, +	-
Epidermal growth factor (EGF)	0, +	+
Transforming growth factor-α (TGF-α)	0, +	+
Transforming growth factor-β (TGF-β)	-, 0 , +	-
Platelet-derived growth factor (PDGF)	0, +	-
Dexamethasone	0, +	0, -
Triiodothyronine	0	0, +
Testosterone	0	-
β-adrenergic agonist	0, +	0, +
Leukaemia inhibitory factor (LIF)	+	?
Retinoic acid	?	+
Linoleic acid	?	+
Calcitonin-gene-related peptide (CGRP)	?	+

Laminin was shown to promote cell adhesion and differentiation of rat and mouse myoblasts [102-105], although the response to laminin was reported to be dependent on the age of the donor rats [103]. Entactin is responsible for long-term maintenance and maturation of contractile rat skeletal myotubes [105]. In a study on chicken myocytes, it was noted that in the absence of collagen, globular syncytial structures were formed instead of long cylindrical myotubes [106]. Another report describes that myogenic chicken cells cultured on type I collagen aligned in parallel, whereas myotubes grown on type V collagen did not have the normal elongated appearance and were not aligned [107], indicating that the correct isoform of collagen is essential for myotube differentiation. But why can human myotubes differentiate so well on uncoated plasticware and even on glass [34, 51, 85, 97]? An explanation might be that in mixed cultures of primary muscle cells, the cultured cells produce basement membrane components themselves. Differentiated mouse myotubes contained increased levels of type IV collagen and laminin as compared to mononuclear myoblasts [108]. Fibroblasts are known to produce collagen IV, which contributes to the surface of myotubes [109]. Furthermore, in contrast to myotubes that differentiate in a mixed cell population, myotubes developed from clonal myoblasts do not seem to form a basement membrane [110-112]. In conclusion, coating of tissue culture surfaces might be extremely important in cultures containing no, or very few fibroblasts (e.g. in clonal cultures or muscle cell-enriched cultures), whereas in mixed cell populations cultured in optimized media, sufficient basement membrane proteins are synthesized by the cells to promote terminal differentiation.

3.3 Enrichment of the Cell Population for Muscle Cells

A frequent problem in primary cell culture is the presence of non-muscle cell types [113], especially fibroblasts, which can overgrow the cell type of interest. Several protocols have been described to remove fibroblasts from primary cultures of human skeletal muscle cells (Table 3). A very simple method is preplating the isolated cells on tissue culture dishes for 20 - 30 minutes preceding the eventual seeding [50, 114]. Fibroblasts adhere faster to the dishes in comparison to satellite cells, and after 20 minutes, the non-adhering cells are replated in a fresh tissue culture dish. The vast majority (96-100%) of these non-adhering cells were reported to be satellite cells [50]. Conversely, fibroblasts seem to be less sensitive to trypsin and adhere longer to the dish surface relative to myoblasts during passage detachment. Keeping trypsin dissociation as brief as possible, fibroblast contamination does not go beyond 5% [115]. The use of a special plastic support that favors myoblast proliferation more than fibroblast proliferation (Primaria, Falcon) can also help to obtain fibroblast-poor myoblast cultures [88].

Table 3. Procedures to enrich the cell population for muscle cells

Procedure	Special requirements	References
Pre-plating	None	50, 114
Myoblast proliferation favoring dishes	Primaria dishes (Falcon)	88
Flow sizing	Flow cytometer	116
FACS	Fluorescence activated cell sorter; 5.1H11 antibody	101
Percoll density gradient centrifugation	Percoll; centrifuge	119, 120, 121

A further, more elaborate protocol for the enrichment of satellite cells is flow cytometry [116]. This method takes advantage of the fact that satellite cells are smaller than most of the other cell types isolated after dissociation of biopsy material, and therefore can be size separated by flow cytometry. Over 98% of these cells were stained by the 5.1H11 monoclonal antibody [117] that labels an NCAM isoform specifically expressed on the surface of myoblasts and myotubes [118].

The 5.1H11 antibody was also used to purify human myoblasts with a fluorescence-activated cell sorter [101]. A cell suspension is incubated with hybridoma supernatant, and subsequently with biotinylated anti-mouse IgG antibody and Texas Red®-avidin. Alternatively, cells were incubated with Pan F haptenated antibody and FITC-conjugated anti-Pan F hapten antibody. The amplification of the fluorescent signal is necessary due to the low density of the antigen on the surface of the myocytes. With this method, approximately 6000 myoblasts per second can be purified from a mixed cell population. An enrichment of myoblasts to greater than 99% was reported [101]. Essential for optimal cell separation is the complete dissociation of cell monolayers in order to avoid cell clumping and selection of 5.1H11 negative cells simply by adherence to myocytes.

Two further methods used Percoll density gradient centrifugation for the purification of chicken [119, 120] and rodent [121] satellite cells. This technique might also be useful to enrich the myoblast fraction from human myoblasts. The protocols take advantage of the fact that in a discontinuous Percoll gradient, satellite cells can be separated from fibroblasts, erythrocytes and fiber debris.

3.4 Passaging and Conservation

In monolayer cultures, skeletal muscle cells will differentiate upon reaching confluence due to contact inhibition, even in high nutrition media. To prevent this, once cells have fused and exhibit irreversible withdrawal from the cell cycle, cells have to be detached from the growth surface by treatment with a trypsin solution, diluted, and plated in a fresh Petri dish. Although trypsinization is a standard procedure described in every culture manual (e.g. refs 53, 122), the optimal procedure depends on how firmly a specific cell type attaches to the surface and how sensitive the cells are to different concentrations of trypsin. Our human skeletal muscle cell cultures are passaged before the cultures reach confluence and before any myotubes are visible. The medium is aspirated from the cells and monolayers are treated with warm 0.5mM EDTA in DPBS for 1 minute. After aspiration of the EDTA solution, a thin layer of pre-warmed 0.025% trypsin (1:250) is added, and the cells are placed at 37°C for a few minutes until they detach from the surface and float. The action of trypsin is stopped by adding an equal volume of culture medium, and the cells are collected by centrifugation (6 min, 50*g*) in a 15 ml conical tube. After careful aspiration of the supernatant, the cell pellet is loosened by gently flicking the tube, the cells are resuspended in culture medium and plated in fresh culture dishes. A maximum subcultivation ratio of 1:6 is recommended. If trypsinized cells are to be frozen, the cell pellet obtained after centrifugation has to be loosened, suspended in sterile freezing medium at a density between 2×10^5 and 5×10^6 cells per ml, and frozen as described above for skeletal muscle tissue fragments (see section 2.2). Freezing of skeletal muscle cells does not significantly alter their capacity to proliferate and differentiate if the frozen cells were not grown for a long period in culture [49, 50, 52]. It is, therefore, advisable to freeze large numbers of vials from the earliest passages of each individual culture.

3.5 Extending the Life Span of Muscle Cells

Since the moderate proliferative capacity of primary human skeletal muscle cells is a limiting factor in studies for which large numbers of myoblasts are required, several groups have tried to immortalize human myoblasts, or at least to extend their life span [123-128]. Transfection of human myoblasts with constructs that carry the gene encoding the SV40 large T antigen resulted in an extended life span of these cells while their capacity to differentiate in a nearly physiological manner was more or less well preserved. Expression of the T antigen has been used to immortalize several cell types of human origin [129]. This immortalization depends on the inactivation of tumor suppressor proteins p53, retinoblastoma gene product (RB), and the RB-related proteins p107 and p130 [130]. RB, in turn, seems to bind and inactivate MyoD and myogenin, thus inhibiting de-differentiation of myotubes [131, 132]. Therefore, in myoblasts the expression of the T antigen has to be switched off in order to allow complete differentiation of the cells [124, 125]. For this purpose, T antigen expression was linked to an inducible promoter or a temperature-sensitive version of T-antigen [123, 126], or to the vimentin promoter that is down-regulated upon differentiation of muscle cells [125, 127]. Expression of the T antigen reduced the doubling time of fetal myoblasts and satellite cells, while the number of mean population doublings was considerably increased. A similar effect was obtained by infecting human satellite cells with viral constructs carrying the E6 and E7 genes of human papillomavirus [128]. Expression of these genes immortalizes cells by a similar mechanism to that described for the large T antigen [133]. Clones of myoblasts expressing the E6 and E7 genes often showed an extended life-span. The continuous expression of the genes resulted, however, in limited muscle differentiation [128].

4. ASSAY TECHNIQUES

Since all available methods to isolate skeletal muscle cells from biopsy material co-purify non-muscle cells, it is extremely important to establish the number of muscle cells within the mixed cell population. Diseased muscles may contain considerable numbers of fibroblasts, especially when the disease is accompanied by a proliferation of interstitial tissue. The large number of fibroblasts may easily overgrow the myocytes, which would necessitate purification of the myocytes by one of the methods described in Section 3.3. Furthermore, myocytes might have problems finding partners to fuse with, resulting in very poor differentiation.

4.1 Morphology

Unfortunately, proliferating myoblasts, early myocytes and fibroblasts have similar morphologies, making identification by light microscopy extremely difficult, even for experienced researchers [52, 114, 134]. Immediately after plating, freshly isolated myoblasts or satellite cells have a rounded appearance that changes gradually following attachment to the substratum. The cells take on a spindle-shaped to polygonal appearance, and as long as the cultures are non-confluent, rounded-up mitotic cells can be observed (Figure 2). Withdrawal of most of the nutrients from the culture medium induces differentiation that results in the development of long, cross-striated and spontaneously contracting myotubes. Usually, this process takes six or seven days using the culture conditions described above. The fusion of myoblasts to myotubes is irreversible but, during all stages of differentiation, numerous mononuclear cells that have withdrawn from the cell cycle without fusing are observed, and these cells can re-enter the cell cycle upon addition of nutrients. Since muscle cells cannot be microscopically identified until they start to differentiate, antibodies specific for myoblast markers must be used. One of the antibodies, 5.1H11 [117], was described above. This antibody can be obtained from the Developmental Studies Hybridoma Bank maintained by the University of Iowa (Iowa City, IA). More generally used, however, are antibodies against desmin, the muscle-specific intermediate filament protein. In contrast to fibroblasts, proliferating human satellite cells are stained by antibodies specific for desmin [52, 135].

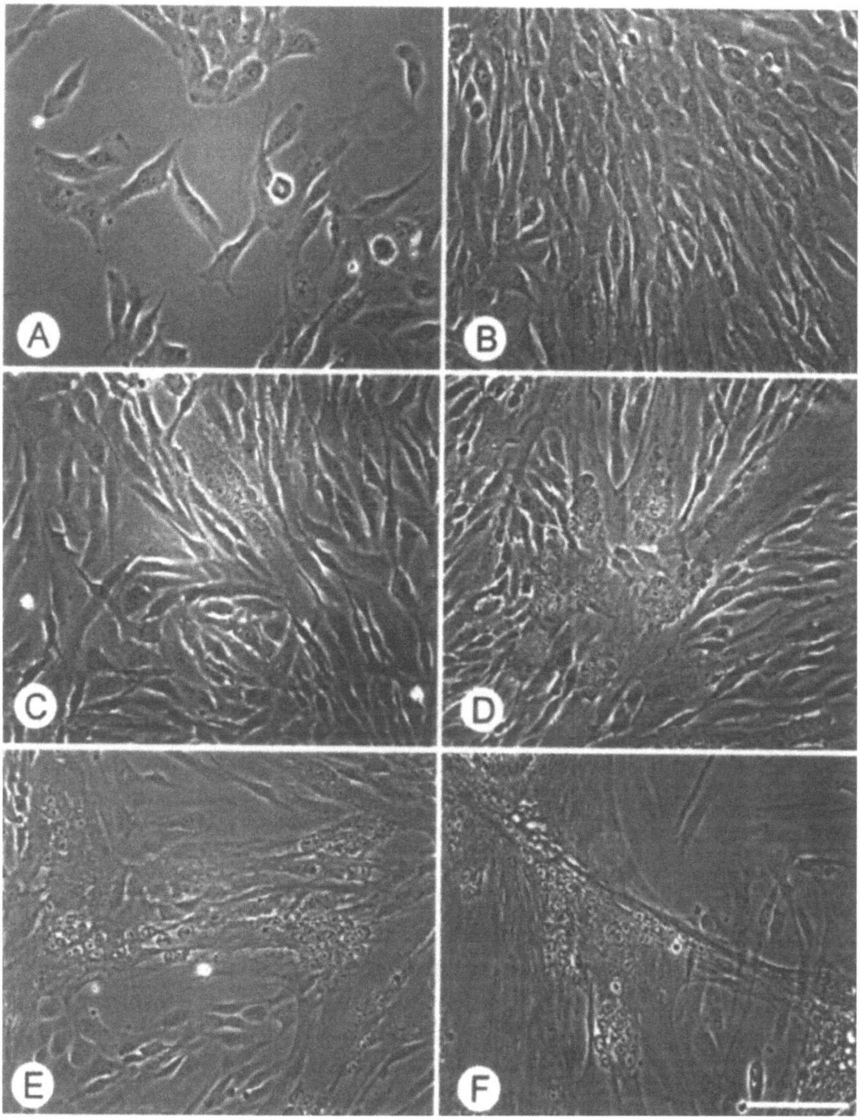

Figure 2. Phase contrast photomicrographs of methanol-fixed cultured human skeletal muscle cells during several stages of differentiation. Proliferating cells are depicted before reaching confluence (A) and upon reaching confluence (B). In these stages, all cells are mononuclear. The rounded-up cells visible in A are mitotic cells. Within one day following induction of differentiation by nutrient-withdrawal, the first small myotubes can be observed (C). The number myotubes, as well as the size of the individual myotubes, increases rapidly during the next few days, and usually branched myotubes can be observed (D-F). After approximately six days of differentiation, the myotubes often contract spontaneously and show a cross-striated appearance (F, see also Fig. 3). Bar: 100 µm.

4.2 Degree of Differentiation

Maturation of cultured myotubes can be studied by various methods. The morphology changes dramatically during development (Figure 3). Early myotubes are relatively thin and short with centrally located, clustered nuclei. More differentiated myotubes are longer and thicker, and nuclei are often more peripheral. With phase-contrast, development of cables (immature myofibrils) can be seen, which occupy most of the sarcoplasm, forcing the nuclei to the subsarcolemmal region. From these immature myofibrils, structures with an obvious cross-striated appearance develop. Eventually, these align and the whole, often spontaneously contracting, myotube displays a myofibrillar cross-striation.

This process can be studied immunocytochemically using antibodies recognizing components such as titin (Figure 4) [52, 85, 99]. The supramolecular organization of this giant protein is a good marker for the development of myofibrils and thus, the differentiation of myotubes [52]. In proliferating myoblasts titin cannot be detected, but in differentiating human skeletal muscle cells, titin is one of the first myofibrillar proteins expressed after the induction of differentiation, even before cell elongation or cell fusion occurs. In these cells, titin is distributed in a punctate pattern. Then these titin aggregates associate with actin bundles, the so-called stress fiber-like structures. During further differentiation, titin is reorganized in longitudinal fibrils that eventually change to cross-striated myofibrils. Terminal differentiation is reached when the previously individual myofibrils lie in register and the myotubes show a mature cross-striated morphology when stained for titin [52, 99].

Biochemical assays include analysis of creatine kinase (CK) isozymes. CK is composed of two subunits (CK-B and/or CK-M). In proliferating myoblasts, only the CK-BB isoform is detected, and in differentiated normal adult muscle, the muscle CK-MM is almost the only isoform present. Intermediate stages contain a mixture of both isoforms together with the heterodimer CK-MB. The assays take advantage of the fact that during embryogenesis, regeneration and *in vitro* differentiation, the levels of the individual CK dimers present in the muscle cells gradually change [114, 136-138]. The percentage of CK-MM is a marker for the maturation grade. The highest percentages of CK-MM reported in aneurally cultured human myotubes and using optimized media were over 60% [51, 139], while in cultures with standard culture media, CK-MM percentages of 15 to 20% are usually reached [51]. The activity of CK and other enzymes such as cytochrome c oxidase, phosphorylase, citrate synthase and AMP deaminase also correlate with the maturation grade, and are useful differentiation markers [51, 89].

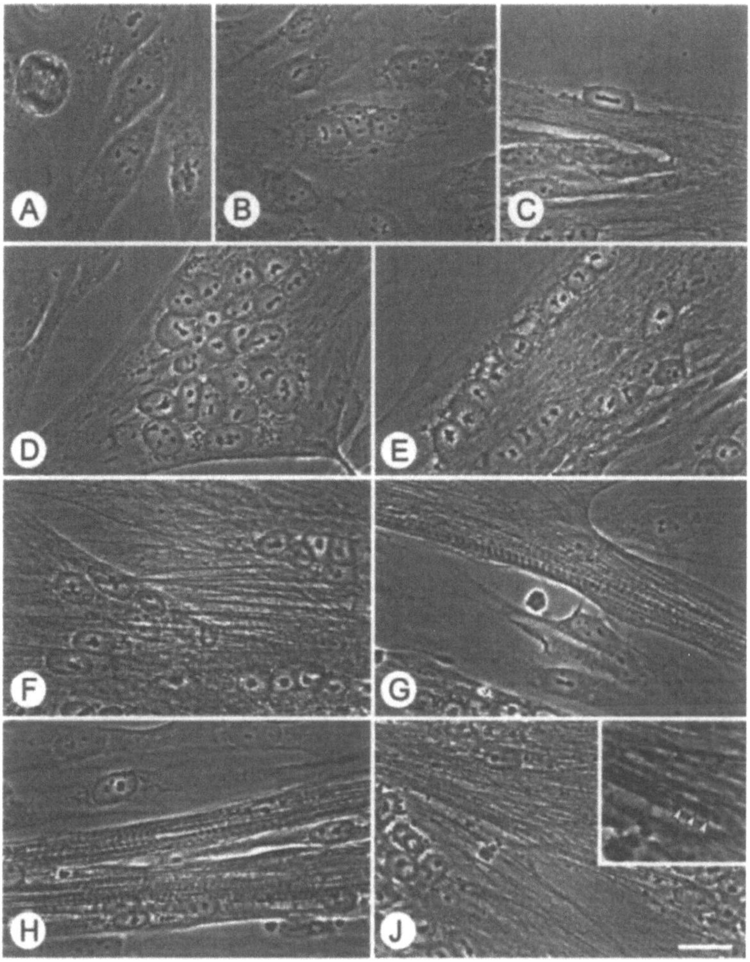

Figure 3. Assembly of myofibrils in phase contrast photomicrographs of methanol-fixed cultured human skeletal muscle cells during several stages of differentiation. In proliferating cell cultures (A), cells are mononuclear and mitotic cells with separated sister chromatids may be observed. Shortly after induction of differentiation, the first myoblasts fuse to form small myotubes (B). In these developmental stages, hardly any filamentous material is observed. During the next few days, mononuclear myoblasts fuse with the early myotubes and large, multinuclear myotubes are observed (C-J). The amount of filamentous material gradually increases. During the first approximately two days of differentiation, this material has a thin and non-structured appearance (C-E), but occupies most of the myotube, forcing the nuclei to the periphery. Thick cables later appear (F) that gradually develop cross-striations (F,G), indicating that myofibrillar structures are assembled (2 - 4 days of differentiation). The sarcoplasm of the most mature myotubes, after approximately 6 days of differentiation (H,J), is filled with large numbers of fully developed myofibrils with discernable A-bands, I-bands, and Z-discs (arrows in J, inset). Bar: 20 μm or 8 μm (inset).

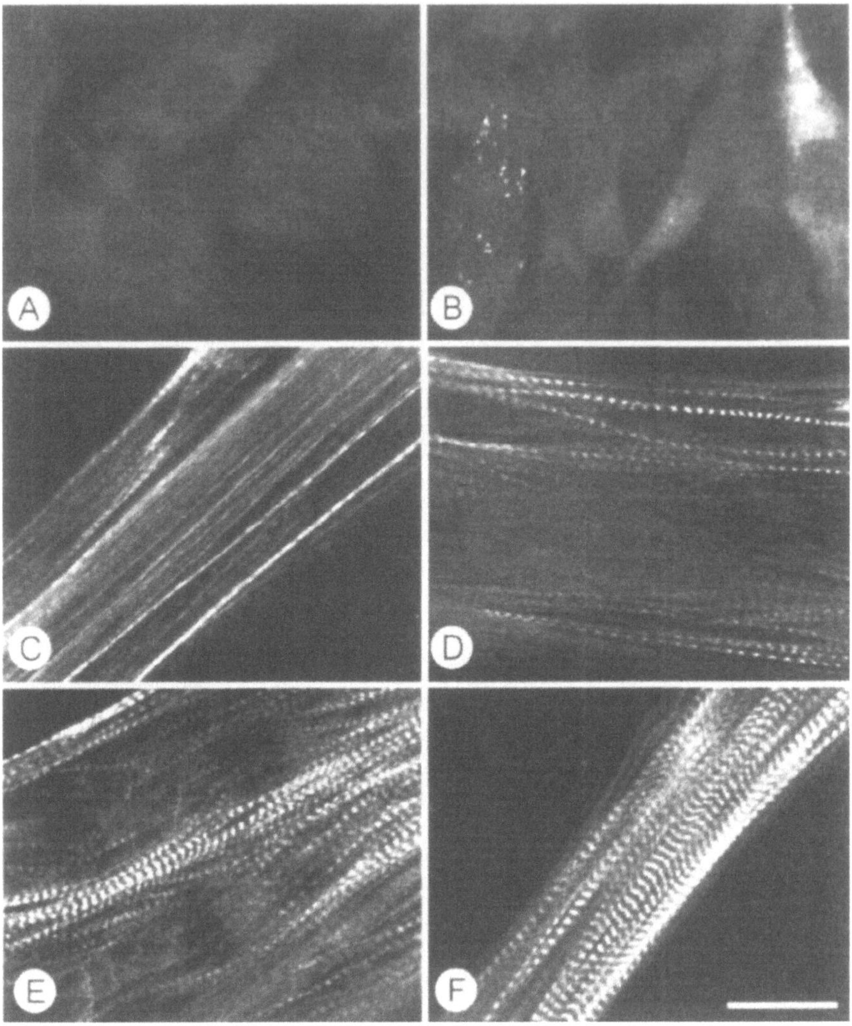

Figure 4. Immunocytochemical analysis of myofibril assembly. Depicted are immunofluorescence micrographs of different stages of methanol/acetone fixed proliferating (A) or differentiating (B-F) human skeletal muscle cells stained with an antibody that recognizes a titin epitope close to the Z-disc. Whereas in proliferating myoblasts titin cannot be detected (A), within 16 hours following the induction of differentiation, titin is expressed in a diffuse or punctate pattern (B). Subsequently, these titin aggregates associate with actin bundles, the so-called stress fiber-like structures, resulting in a continuous staining of these structures (C). During further differentiation, titin is reorganized and observed as regularly spaced dots (D) that eventually change into cross-striations (E). In mature myotubes, individual myofibrils register and the complete myotube shows a cross-striated morphology when stained for titin (F). Bar: 25 μm.

Electrophysiological methods to estimate maturity of muscle cultures include the analysis of resting membrane potentials [95, 140-142]. Dependent on the differentiation grade of the myotubes, membrane potentials may achieve values close or equal to those of around -60 to -80 mV reported for adult human muscle [140, 143].

4.3 Enhancement of Differentiation by *In Vitro* Innervation

Although improved media have been developed, cultured human skeletal muscle cells remain in a relatively immature state not only in comparison to adult myofibers *in vivo*, but also to cultured primary myotubes from other species. The knowledge that terminal myofiber differentiation *in vivo* is only reached upon innervation stimulated several investigators to develop nerve-muscle co-cultures. Early experiments were performed with embryonic rodent spinal cord explants and human or rodent skeletal muscle fragments placed on collagen-coated glass coverslips in culture medium containing high levels of human serum and rat embryo extract. These experiments showed that myofibers within the muscle fragments are innervated by axonal outgrowths of nerve cells from the spinal cord explant. Neuromuscular connections had a well-differentiated ultrastructure and neuromuscular transmission took place, and innervation had dramatic stimulatory effects on regenerative capacity and survival of the muscle fibers [69, 144-147]. This organ culture technique requires considerable experience in manipulating the tissues and maintaining the cultures, but allows the evaluation of regeneration and innervation in small pieces of muscle tissue. However, due to the minimal amount of available material, organ cultures may be less well suited for biochemical studies, although analyses of enzyme profiles has been reported [147].

Completely *de novo* formation of specialized sites on the surface of myotubes that will further differentiate to form neuromuscular junctions can be studied by innervation of monolayer cultures. In the organ culture system, new myotubes develop within existing basal lamina that are often re-innervated at partly degenerated original synaptic sites. In contrast, in monolayer cultures, all basal lamina structures including synaptic sites are assembled *de novo*. This allows the analysis of consecutive developmental stages during the maturation of neuromuscular junctions. Pioneering work with dissociated suspensions of chicken myoblasts and embryonic motor neurons was described by Shimada and co-workers [148-152], while a method to innervate monolayers of human skeletal muscle cells was developed in the laboratory of Askanas [153]. In this procedure, muscle cells obtained using the explant-reexplant technique are allowed to grow and differentiate in a medium containing F14 medium [56] with 6000 mg/l glucose and supplemented with 10% FCS, 25 ng/ml FGF, 10 ng/ml EGF, and 10 μg/ml insulin [35]. Approximately 10 to 15 days after the appearance of the first myotubes, four transverse slices of spinal cord with attached dorsal root ganglia [154] from 12 to 14 day-old rat embryos are placed in each 35mm Petri dish with myotubes. The co-cultures are kept in the same medium from which FGF and EGF are omitted. Within 2 to 3 days of explantation of the neural tissue, contacts between neurites and myotubes develop. Whereas non-innervated myotubes degenerate 6 to 8 weeks after the start of co-culture, in the same Petri dish cross-striated and contracting innervated myotubes can be maintained for months. When compared to non-innervated cultures, these innervated cultures: i) contain more continuously contracting myotubes [153], ii) express increased levels of muscle-specific isoforms of several enzymes [155, 156], iii) have higher resting membrane potentials [157] and iv) contain myotubes with a more mature distribution of basement membrane components and T-tubules [153, 158].

It must be noted that the differentiation-stimulating effects of innervation, at least in part, are mimicked by optimized media containing extracts from neonatal rat brain [51, 97] or extracts from peripheral nerves [159-161].

4.4 Gene Delivery to Cultured Muscle Cells

Although several reports have been published that describe transfer of genes cloned in plasmid-based eukaryotic expression vectors to cultured chicken [162], quail [163, 164], mouse [165-167] or rat primary skeletal muscle cells, high transfection efficiencies are difficult to obtain. Probably the lowest transfection levels were found in human myocytes: very low transfection efficiencies within the range of 0 - 1% were obtained [125, 168]. This urged many researchers to use recombinant viruses to introduce genes into mammalian myocytes [126, 128, 169-176]. The disadvantage of this method is that retroviruses often are highly cytotoxic. Currently, novel recombinant viral vectors and transfection protocols are being developed that might optimize infection efficiencies in human muscle cells [128, 170, 173-177].

4.5 Making Your Own Muscle Cells

In cases where no skeletal muscle biopsies are available, other cell types that are much easier to obtain, such as fibroblasts, amniocytes and chorionic-villus cells may be forced into myogenesis by the introduction of MyoD [3]. These converted cells were shown to express skeletal muscle specific proteins, to be able to fuse, to develop mature myofibrils, and to express dystrophin [178, 179], enabling analysis of early stages of muscle differentiation by immunohistochemical staining. Furthermore, expression of muscle-specific mRNAs in these cells facilitates RNA-based mutation detection. The limitations of this method in early studies with a retroviral vector (low transduction efficiencies and the necessity to select transduced cells) were overcome by the development of an adenoviral vector carrying MyoD, enabling transduction efficiencies of up to 95% [177, 180].

4.6 Human Skeletal Muscle Cell Lines

Primary skeletal muscle cells can be propagated for several generations. After a number of passages, cell senescence can be anticipated and the cells lose their myogenic phenotype, which might be caused by a progressive telomere shortening [72]. It is therefore important to freeze as many cells as possible from early passages. An alternative would be the establishment of a permanent myogenic cell line. Several rodent myoblast or myoblast-like cell lines have been described that were obtained after cloning primary myogenic cells. Currently, continuous cell lines developed from mouse [74, 181, 182] and rat [183] skeletal muscle are available. Although these cells can fuse into myotubes and are used extensively to study several aspects of muscle differentiation, these cells must be considered as tumor cells and are no longer normal muscle cells. Furthermore, several cell lines have been described that have at least some characteristics of muscle cells. In all of these cell lines, one or more of the myogenic transcription factors is not expressed and terminal differentiation of the cells is inhibited [184].

The generation of human muscle cell clones was described by two groups [185, 186]. Cell lines were derived from individual cells by clonal growth of primary skeletal muscle cells and selection for myogenic potential. The G6 cell line differentiates rapidly and over 60% of the nuclei are found in myotubes after two days of differentiation in a low nutrition medium [185]. Although several aspects of muscle cell differentiation can be examined using these cells [187-190], the fact that they lose a large part of their differentiation potential within three weeks of culture [185] is a major drawback. The RCMH cell line [186] has retained over 200 passages some characteristics of muscle cells, but little fusion occurs [192] and the suitability of this cell line for muscle differentiation studies seems rather limited.

5. CLINICAL USE OF MYOBLASTS

5.1 Transplantation of Myoblasts

Following the discovery that mutations in the dystrophin gene resulted in Duchenne and Becker muscular dystrophy, several groups tried to use gene therapy by transplantation of normal or genetically engineered myoblasts. Fusion of the transplanted myoblasts with the damaged myofibers would restore the expression of dystrophin in these myofibers, and lead to a cessation of muscle fiber necrosis and improved strength of the implanted muscles. Initially, most experiments were performed with mdx mice, an animal model of DMD. In most cases C2 or C2C12 myoblasts, two well characterized mouse myoblast cell lines, were used in such studies. Although expression of wild type dystrophin could be detected in individual diseased muscle fibers upon transplantation, the graft success was low in the absence of immunosuppression. The success rate was barely increased following irradiation of the transplanted muscle to prevent activation of host satellite cells, or pre-conditioning the muscles to be transplanted by damaging them (by treatment with e.g. notexin).

In clinical studies with DMD patients, considerable success was reported by Law and co-workers, with expression of dystrophin in transplanted muscles even 6 years after transplantation [193-195]. These results could, however, not be reproduced by several other groups [196-200]. Within half a year following myoblast transplantation, most patients did not show any expression of donor-derived dystrophin, and in other patients very few muscle fibers showed expression of wild-type dystrophin, leading to a failure to improve muscle strength. The damaging treatments used to improve transplantation efficiency in mice cannot be used in human clinical studies. Therefore, alternative methods to increase transplantation efficiency, such as preincubation of the myoblasts to be transplanted with concanavalin A (to increase the capacity to migrate in the grafted muscle [201]), are being tested. Furthermore, improved methods of cell delivery and acceptable methods to pre-treat the muscles to be grafted are currently being developed.

5.2 Myoblasts and Gene Therapy

The most promising approach is probably the transplantation of autologous, genetically-modified myoblasts. The advantage of this method is that immune responses to the transplanted myoblasts are prevented by the use of primary muscle cell cultures derived from the patient to be treated. These diseased muscle cells are modified by adenoviral dystrophin gene transfer. Since the capacity of these replication-deficient viral constructs is limited, the complete dystrophin cannot be cloned. Instead, a truncated dystrophin (mini-dystrophin) is expressed to convert the DMD phenotype to the less-severe BMD phenotype [202]. Remaining problems include the modest efficiency of myoblast transduction and the loss of viral copy number during replication of the myoblasts. This system might also be useful for delivery of 'normal' proteins and restoration of the diseased phenotype in other diseases with characterized causes. *In vitro* evidence for the functionality of this method was provided by transduction of C2C12 myotubes with an adenoviral vector containing the muscle glycogen phosphorylase cDNA, resulting in synthesis of the enzyme in the myotubes [203]. *In vivo* transduction of the virus to muscle fibers of patients with McArdle's disease, which is caused by a glycogen phosphorylase deficiency, might prove to be of clinical use. The success of methods using recombinant replication-deficient adenoviruses (and plasmids) that drive expression of a cloned cDNA rely on the fact that muscle fibers seem to be unable to eliminate episomal DNA.

The transplantation of genetically-engineered myoblasts, or the transduction of muscle fibers, provides the possibility of therapy for non-muscle diseases. The transplantation of C2C12 cells engineered to express insulin in diabetic mice, for example, resulted in the development of insulin producing and secreting myofibers [204]. These mice showed significantly increased insulinemia and decreased hyperglycemia.

5.3 Myoblasts and Treatment of Heart Failure

The implantation of skeletal myoblasts into the heart after a myocardial infarct seems to be a further promising possibility for clinical use [205-210]. Since cardiomyocytes cannot regenerate, the loss of injured myocardial tissue is irreversible and the wounds are healed by replacement with scar tissue. Among the cell types that could be transplanted to injured hearts in order to replace scar tissue or to prevent scarring, skeletal myoblasts are the cells of choice for several reasons. Although cardiomyocytes appear to be more eligible, the fact that they, in contrast to skeletal myoblasts, do not proliferate in culture makes their use impractical. Skeletal myoblasts have the advantage that high numbers of autologous donor myoblasts can be cultured, allowing reparation of myocardial infarcts in humans. The implanted cells differentiate into fully developed, slow twitch fibers that not only prevent the formation of scar tissue, but also have the capacity to perform cardiac work. This approach has been tested on rats [208, 209], mice [205], and dogs [206, 207] and is a promising novel strategy for the future treatment of heart failure.

6. CONCLUDING REMARKS

The ability of cultured skeletal muscle cells to differentiate makes them an invaluable tool to study early stages of muscle fiber maturation. Proliferating cells that withdraw from the cell cycle, fuse to form multinuclear cells that gradually mature, culminating in the development of spontaneously contracting, cross-striated myotubes. Thus myogenesis or muscle regeneration is recapitulated, enabling *in vitro* analysis of these processes. Most of our current knowledge on how numerous structural proteins are assembled into the extremely ordered myofibrils originates from studies on muscle cell cultures, and a comparison of gene expression patterns in differentiated and non-differentiated muscle cells led to the discovery of the myogenic determination factors. These cultures are unlikely to reach a level of differentiation equal to the in vivo situation, but the identification of hormones and growth factors that stimulate myotube differentiation, and the analysis of the effects of extracellular matrix proteins on differentiation, enabled the development of optimal culture media and substrates. Furthermore, the innervation of human myotubes in nerve-muscle co-cultures of human skeletal muscle cells with rodent spinal cord provides a further opportunity to increase the maturation grade. As to the clinical use of skeletal muscle cell cultures, cultured diseased muscle cells might be used to identify defects causing or accompanying the disease, and to search for novel therapeutic approaches. Transplantation of normal or genetically-engineered myoblasts might be used to compensate for genetic muscle defects or myocardial injuries. Equally promising are the attempts to transplant genetically-engineered myoblasts for therapy of non-muscle diseases, such as diabetes. The steadily increasing number of reports describing experiments with cultures of skeletal muscle cells points to a general acceptance of the culture system as a model for myodifferentiation and as a promising therapeutic tool.

Acknowledgement: The author expresses his special thanks to Dr Dieter Fürst for continuing support, interest, and helpful discussions.

Abbreviations: bHLH: basic helix-loop-helix; BSA: bovine serum albumin; CK: creatine kinase; DMEM : Dulbecco's modified Eagle's medium; DMSO: dimethyl sulfoxide; DPBS: Dulbecco's modified phosphate-buffered saline; (E)MEM: Eagle's minimum essential medium; EDTA: ethylenediaminetetraacetic acid; EGF: epidermal growth factor; FACS: Fluorescence activated cell sorter; FCS: fetal calf serum; FGF: fibroblast growth factor; FITC: fluorescein isothiocyanate; Ig: immunoglobulin; IGF: insulin-like growth factor; NCAM: neural cell adhesion molecule.

REFERENCES

1. Hauschka SD (1994) The embryonic origin of muscle, in AG Engel and C Franzini-Armstrong (eds.), *Myology*. McGraw-Hill, New York, pp. 3-73
2. Kalcheim C, Cinnamon Y, and Kahane N (1999) Myotome formation: a multistage process. *Cell Tissue Res.* 296:161-173
3. Davis RL, Weintraub H, and Lassar AB (1987) Expression of a single transfected cDNA converts fibroblasts to myoblasts. *Cell* 51:987-1000
4. Braun T, Buschhausen-Denker G, Bober E, Tannich E, and Arnold HH (1989) A novel human muscle factor related to but distinct from MyoD1 induces myogenic conversion in 10T1/2 fibroblasts. *EMBO J.* 8:701-709
5. Wright WE, Sassoon DA, and Lin VK (1989) Myogenin, a factor regulating myogenesis, has a domain homologous to MyoD. *Cell* 56:607-617
6. Edmondson DG, and Olson EN (1990) A gene with homology to the myc similarity region of MyoD1 is expressed during myogenesis and is sufficient to activate the muscle differentiation program. *Genes Dev.* 4:1450
7. Rhodes SJ, and Konieczny SF (1989) Identification of MRF4: a new member of the muscle regulatory factor gene family. *Genes Dev.* 3:2050-2061
8. Braun T, Bober E, Winter B, Rosenthal N, and Arnold HH (1990) Myf-6, a new member of the human gene family of myogenic determination factors: evidence for a gene cluster on chromosome 12. *EMBO J.* 9:821-831
9. Miner JH, and Wold B (1990) Herculin, a fourth member of the MyoD family of myogenic regulatory genes. *Proc.Natl.Acad.Sci.U.S.A.* 87:1089-1093
10. Rudnicki MA, Braun T, Hinuma S, and Jaenisch R (1992) Inactivation of MyoD in mice leads to up-regulation of the myogenic HLH gene Myf-5 and results in apparently normal muscle development. *Cell* 71:383-390
11. Braun T, Rudnicki MA, Arnold HH, and Jaenisch R (1992) Targeted inactivation of the muscle regulatory gene Myf-5 results in abnormal rib development and perinatal death. *Cell* 71:369-382
12. Hasty P, Bradley A, Morris JH, Edmondson DG, Venuti JM, Olson EN, and Klein WH (1993) Muscle deficiency and neonatal death in mice with a targeted mutation in the myogenin gene. *Nature* 364:501-506
13. Nabeshima Y, Hanaoka K, Hayasaka M, Esumi E, Li S, and Nonaka I (1993) Myogenin gene disruption results in perinatal lethality because of severe muscle defect. *Nature* 364:532-535
14. Zhang W, Behringer RR, and Olson EN (1995) Inactivation of the myogenic bHLH gene MRF4 results in up-regulation of myogenin and rib anomalies. *Genes Dev.* 9:1388-1399
15. Olson EN, Arnold HH, Rigby PW, and Wold BJ (1996) Know your neighbors: three phenotypes in null mutants of the myogenic bHLH gene MRF4. *Cell* 85:1-4
16. Yoon JK, Olson EN, Arnold HH, and Wold BJ (1997) Different MRF4 knockout alleles differentially disrupt Myf-5 expression: cis-regulatory interactions at the MRF4/Myf-5 locus. *Dev.Biol.* 188:349-362
17. Ludolph DC, and Konieczny SF (1995) Transcription factor families: muscling in on the myogenic program. *FASEB J.* 9:1595-1604
18. Arnold HH, and Braun T (1996) Targeted inactivation of myogenic factor genes reveals their role during mouse myogenesis: a review. *Int.J.Dev.Biol.* 40:345-353
19. Rawls A, and Olson EN (1997) MyoD meets its maker. *Cell* 89:5-8

20. Kablar B, Krastel K, Ying C, Asakura A, Tapscott SJ, and Rudnicki MA (1997) MyoD and Myf-5 differentially regulate the development of limb versus trunk skeletal muscle. *Development* 124:4729-4738
21. Tajbakhsh S, Rocancourt D, Cossu G, and Buckingham M (1997) Redefining the genetic hierarchies controlling skeletal myogenesis: Pax- 3 and Myf-5 act upstream of MyoD. *Cell* 89:127-138
22. Brand-Saberi B, and Christ B (1999) Genetic and epigenetic control of muscle development in vertebrates. *Cell Tissue Res.* 296:199-212
23. Chen JC, and Goldhamer DJ (1999) Transcriptional mechanisms regulating MyoD expression in the mouse. *Cell Tissue Res.* 296:213-219
24. Burkin DJ, and Kaufman SJ (1999) The alpha7beta1 integrin in muscle development and disease. *Cell Tissue Res.* 296:183-190
25. Kaufmann U, Martin B, Link D, Witt K, Zeitler R, Reinhard S, and Starzinski-Powitz A (1999) M-cadherin and its sisters in development of striated muscle. *Cell Tissue Res.* 296:191-198
26. Mauro A (1961) Satellite cells of skeletal muscle fibers. *J.Biophys.Biochem.Cytol.* 9:493-495
27. Katz B (1961) The terminations of the afferent nerve fibre in the muscle spindle of the frog. *Philos.Trans.R.Soc.Lond.* 243:221-240
28. Campion DR (1984) The muscle satellite cell: a review. *Int.Rev.Cytol.* 87:225-51:225-251
29. Grounds MD (1991) Towards understanding skeletal muscle regeneration. *Pathol.Res.Pract.* 187:1-22
30. Bischoff R (1994) The satellite cell and muscle regeneration, in AG Engel and C Franzini-Armstrong (eds.), *Myology*, McGraw-Hill, New York, pp. 97-118
31. Maltin CA, Harris JB, and Cullen MJ (1983) Regeneration of mammalian skeletal muscle following the injection of the snake-venom toxin, taipoxin. *Cell Tissue Res.*232:565-577
32. Schultz E, Jaryszak DL, and Valliere CR (1985) Response of satellite cells to focal skeletal muscle injury. *Muscle Nerve* 8:217-222
33. Morgan JE, Coulton GR, and Partridge TA (1987) Muscle precursor cells invade and repopulate freeze-killed muscles. *J. Muscle Res.Cell Motil.*8:386-396
34. Ham RG, St.Clair JA, Webster C, and Blau HM (1988) Improved media for normal human muscle satellite cells: serum-free clonal growth and enhanced growth with low serum. *In Vitro Cell Dev.Biol.* 24:833-844
35. Askanas V, and Gallez-Hawkins G (1985) Synergistic influence of polypeptide growth factors on cultured human muscle. *Arch.Neurol.* 42:749-752
36. Husmann I, Soulet L, Gautron J, Martelly I, and Barritault D (1996) Growth factors in skeletal muscle regeneration. *Cytokine Growth Factor Rev.* 7:249-258
37. Brameld JM, Buttery PJ, Dawson JM, and Harper JM (1998) Nutritional and hormonal control of skeletal-muscle cell growth and differentiation. *Proc.Nutr.Soc.* 57:207-217
38. Semsarian C, Sutrave P, Richmond DR, and Graham RM (1999) Insulin-like growth factor (IGF-I) induces myotube hypertrophy associated with an increase in anaerobic glycolysis in a clonal skeletal-muscle cell model. *Biochem.J.*339:443-451
39. Adams GR, and McCue SA (1998) Localized infusion of IGF-I results in skeletal muscle hypertrophy in rats. *J.Appl.Physiol.* 84:1716-1722
40. Coleman ME, DeMayo F, Yin KC, Lee HM, Geske R, Montgomery C, and Schwartz RJ (1995) Myogenic vector expression of insulin-like growth factor I stimulates muscle cell differentiation and myofiber hypertrophy in transgenic mice. *J.Biol.Chem.* 270:12109-12116

41. Musaro A, and Rosenthal N (1999) Maturation of the myogenic program is induced by postmitotic expression of insulin-like growth factor I. *Mol.Cell Biol.* 19:3115-3124
42. Musaro A, McCullagh KJ, Naya FJ, Olson EN, and Rosenthal N (1999) IGF-1 induces skeletal myocyte hypertrophy through calcineurin in association with GATA-2 and NF-ATc1. *Nature* 400:581-585
43. Semsarian C, Wu MJ, Ju YK, Marciniec T, Yeoh T, Allen DG, Harvey RP, and Graham RM (1999) Skeletal muscle hypertrophy is mediated by a Ca2+-dependent calcineurin signalling pathway. *Nature* 400:576-581
44. Bonavaud S, Thibert P, Gherardi RK, and Barlovatz-Meimon G (1997) Primary human muscle satellite cell culture: variations of cell yield, proliferation and differentiation rates according to age and sex of donors, site of muscle biopsy, and delay before processing. *Biol.Cell* 89:233-240
45. Partridge TA (1997) Tissue culture of skeletal muscle. *Methods Mol.Biol.* 75:131-144
46. Jiguo Y, Gautel M, Price MG, Robson R, Wiche G, Thornell L-E (1999) Cytoskeletal changes related to mucle soreness revisited. *J Muscle Res Cell Motil* 129:108(Abstract)
47. Yasin R, Van Beers G, Nurse KC, Al-Ani S, Landon DN, and Thompson EJ (1977) A quantitative technique for growing human adult skeletal muscle in culture starting from mononucleated cells. *J.Neurol.Sci.* 32:347-360
48. Askanas V, and Engel WK (1975) A new program for investigating adult human skeletal muscle grown aneurally in tissue culture. *Neurology* 25:58-67
49. Carter LS, and Askanas V (1981) Vital-freezing of human muscle cultures for storage and reculture. *Muscle Nerve* 4:367-369
50. Blau HM, and Webster C (1981) Isolation and characterization of human muscle cells. *Proc.Natl.Acad.Sci.U.S.A.* 78:5623-5627
51. Benders AA, van Kuppevelt TH, Oosterhof A, and Veerkamp JH (1991) The biochemical and structural maturation of human skeletal muscle cells in culture: the effect of the serum substitute Ultroser G. *Exp.Cell Res.* 195:284-294
52. van der Ven PFM, Schaart G, Jap PHK, Sengers RCA, Stadhouders AM, and Ramaekers FCS (1992) Differentiation of human skeletal muscle cells in culture: maturation as indicated by titin and desmin striation. *Cell Tissue Res.* 270:189-198
53. Freshney RI (1994) *Culture of animal cells: a manual of basic technique*, 3rd edn. Wiley-Liss, Inc., New York
54. Cossu G, Zani B, Coletta M, Bouche M, Pacifici M, and Molinaro M (1980) In vitro differentiation of satellite cells isolated from normal and dystrophic mammalian muscles. A comparison with embryonic myogenic cells. *Cell Differ.* 9:357-368
55. Gibson MC, and Schultz E (1983) Age-related differences in absolute numbers of skeletal muscle satellite cells. *Muscle Nerve* 6:574-580
56. Vogel Z, Sytkowski AJ, and Nirenberg MW (1972) Acetylcholine receptors of muscle grown in vitro. *Proc.Natl.Acad.Sci.U.S.A.* 69:3180-3184
57. Askanas V, Cave S, Gallez-Hawkins G, and Engel WK (1985) Fibroblast growth factor, epidermal growth factor and insulin exert a neuronal-like influence on acetylcholine receptors in aneurally cultured human muscle. *Neurosci.Lett.* 61:213-219
58. Kobayashi T, Askanas V, Saito K, Engel WK, and Ishikawa K (1990) Abnormalities of aneural and innervated cultured muscle fibers from patients with myotonic atrophy (dystrophy). *Arch.Neurol.* 47:893-896
59. Witkowski JA, Durbidge M, and Dubowitz V (1976) Growth of human muscle in tissue culture. An improved technique. *In Vitro* 12:98-106
60. Bischoff R (1986) Proliferation of muscle satellite cells on intact myofibers in culture. *Dev.Biol.* 115:129-139

61. Bekoff A, and Betz W (1977) Properties of isolated adult rat muscle fibres maintained in tissue culture. *J.Physiol.(Lond.)* 271:537-547
62. Bischoff R (1990) Interaction between satellite cells and skeletal muscle fibers. *Development* 109:943-952
63. Bischoff R (1989) Analysis of muscle regeneration using single myofibers in culture. *Med.Sci.Sports Exerc.* 21:S164-S172
64. Bischoff R (1990) Cell cycle commitment of rat muscle satellite cells. *J.Cell Biol.* 111:201-207
65. Bischoff R (1986) A satellite cell mitogen from crushed adult muscle. *Dev.Biol.* 115:140-147
66. Rosenblatt JD, Lunt AI, Parry DJ, and Partridge TA (1995) Culturing satellite cells from living single muscle fiber explants. *In Vitro Cell Dev.Biol.Anim.* 31:773-779
67. Rosenblatt JD, Parry DJ, and Partridge TA (1996) Phenotype of adult mouse muscle myoblasts reflects their fiber type of origin. *Differentiation* 60:39-45
68. Zuurveld JGEM (1984) *Skeletal muscle. Cellular biochemistry of developing and differentiated myofibers.* Thesis. University of Nijmegen, Nijmegen
69. Peterson ER, Crain SM (1979) Maturation of human muscle after innervation by fetal mouse spinal cord explants in long-term cultures, in A Mauro (ed.), *Muscle Regeneration.* Raven Press, New York, pp. 429-441
70. Ecob-Prince MS, and Brown AE (1988) Morphological differentiation of human muscle cocultured with mouse spinal cord. *J.Neurol.Sci.* 83:179-190
71. Ecob-Prince MS, and Cullen MJ (1988) Atypical persisting fibres in explants of human muscle cocultured with embryonic nerve cells. *J.Neurol.Sci.* 83:321-333
72. Decary S, Mouly V, Hamida CB, Sautet A, Barbet JP, and Butler BG (1997) Replicative potential and telomere length in human skeletal muscle: implications for satellite cell-mediated gene therapy. *Hum.Gene Ther.* 8:1429-1438
73. Schultz E, and Lipton BH (1982) Skeletal muscle satellite cells: changes in proliferation potential as a function of age. *Mech.Ageing Dev.* 20:377-383
74. Hauschka SD, Linkhart TA, Clegg C, and Merrill G (1979) Clonal studies of human and mouse muscle, in A Mauro (ed.), *Muscle Regeneration.* Raven Press, New York, pp. 311-322
75. Yasin R, Kundu D, and Thompson EJ (1981) Growth of adult human cells in culture at clonal densities. *Cell Differ.* 10:131-137
76. Schultz E (1989) Satellite cell behavior during skeletal muscle growth and regeneration. *Med.Sci.Sports Exerc.* 21:S181-S186
77. Feldman JL, and Stockdale FE (1991) Skeletal muscle satellite cell diversity: satellite cells form fibers of different types in cell culture. *Dev.Biol.* 143:320-334
78. Hartley RS, Bandman E, and Yablonka-Reuveni Z (1991) Myoblasts from fetal and adult skeletal muscle regulate myosin expression differently. *Dev.Biol.* 148:249-260
79. Barjot C, Cotten ML, Goblet C, Whalen RG, and Bacou F (1995) Expression of myosin heavy chain and of myogenic regulatory factor genes in fast or slow rabbit muscle satellite cell cultures. *J.Muscle Res.Cell Motil.*16:619-628
80. Ghosh S, and Dhoot GK (1998) Evidence for distinct fast and slow myogenic cell lineages in human foetal skeletal muscle. *J.Muscle Res.Cell Motil.* 19:431-441
81. Dusterhoft S, Yablonka-Reuveni Z, and Pette D (1990) Characterization of myosin isoforms in satellite cell cultures from adult rat diaphragm, soleus and tibialis anterior muscles. *Differentiation* 45:185-191
82. Cho M, Webster SG, and Blau HM (1993) Evidence for myoblast-extrinsic regulation of slow myosin heavy chain expression during muscle fiber formation in embryonic development. *J.Cell Biol.* 121:795-810

83. Edom F, Mouly V, Barbet JP, Fiszman MY, and Butler-Browne GS (1994) Clones of human satellite cells can express in vitro both fast and slow myosin heavy chains. *Dev.Biol.* 164:219-229
84. Dusterhoft S, and Pette D (1993) Satellite cells from slow rat muscle express slow myosin under appropriate culture conditions. *Differentiation* 53:25-33
85. van der Ven PFM, and Fürst DO (1997) Assembly of titin, myomesin and M-protein into the sarcomeric M band in differentiating human skeletal muscle cells in vitro. *Cell Struct.Funct.* 22:163-171
86. Mues A, van der Ven PFM, Young P, Fürst DO, and Gautel M (1998) Two immunoglobulin-like domains of the Z-disc portion of titin interact in a conformation-dependent way with telethonin. *FEBS Lett.* 428:111-114
87. van der Ven PFM, Ehler E, Perriard JC, and Fürst DO (1999) Thick filament assembly occurs after the formation of a cytoskeletal scaffold. *J.Muscle Res.Cell Motil.* 20:569-579
88. Delaporte C, Dautreaux B, and Fardeau M (1986) Human myotube differentiation in vitro in different culture conditions. *Biol.Cell* 57:17-22
89. Zuurveld JGEM, Oosterhof A, Veerkamp JH, and van Moerkerk HT (1985) Oxidative metabolism of cultured human skeletal muscle cells in comparison with biopsy material. *Biochim.Biophys.Acta* 844:1-8
90. Mendell JR, Roelofs RI, and Engel WK (1972) Ultrastructural development of explanted human skeletal muscle in tissue culture. *J.Neuropathol.Exp.Neurol.* 31:433-436
91. Ham RG (1963) An improved nutrient solution for diploid Chinese hamster and human cell lines. *Exp.Cell Res.* 29:515-526
92. Hauschka SD (1974) Clonal analysis of vertebrate myogenesis. II. Environmental influences upon human muscle differentiation. *Dev.Biol.* 37:329-344
93. Baroffio A, Bochaton-Piallat ML, Gabbiani G, and Bader CR (1995) Heterogeneity in the progeny of single human muscle satellite cells. *Differentiation* 59:259-268
94. Bishop A, Gallup B, Skeate Y, and Dubowitz V (1971) Morphological studies on normal and diseased human muscle in culture. *J.Neurol.Sci.* 13:333-350
95. Harvey AL, Robertson JG, and Witkowski JA (1979) Maturation of human skeletal muscle fibres in explant tissue culture. *J.Neurol.Sci.* 41:115-122
96. Kameda N, Ueda H, Ohno S, Shimokawa M, Usuki F, Ishiura S, and Kobayashi T (1998) Developmental regulation of myotonic dystrophy protein kinase in human muscle cells in vitro. *Neuroscience* 85:311-322
97. van Kuppevelt THMSM, Benders AAGM, Versteeg EM, and Veerkamp JH (1992) Ultroser G and brain extract induce a continuous basement membrane with specific synaptic elements in aneurally cultured human skeletal muscle cells. *Exp.Cell Res.* 200:306-315
98. St.Clair JA, Meyer-Demarest SD, and Ham RG (1992) Improved medium with EGF and BSA for differentiated human skeletal muscle cells. *Muscle Nerve* 15:774-779
99. van der Ven PFM, Schaart G, Croes HJ, Jap PHK, Ginsel LA, and Ramaekers FCS (1993) Titin aggregates associated with intermediate filaments align along stress fiber-like structures during human skeletal muscle cell differentiation. *J.Cell Sci.* 106:749-759
100. van der Ven PFM, Jap PHK, Barth PG, Sengers RCA, Ramaekers FCS, and Stadhouders AM (1995) Abnormal expression of intermediate filament proteins in X-linked myotubular myopathy is not reproduced in vitro. *Neuromuscul.Disord.* 5:267-275
101. Webster C, Pavlath GK, Parks DR, Walsh FS, and Blau HM (1988) Isolation of human myoblasts with the fluorescence-activated cell sorter. *Exp.Cell Res.* 174:252-265

102. Kuhl U, Öcalan M, Timpl R, and von der Mark K (1986) Role of laminin and fibronectin in selecting myogenic versus fibrogenic cells from skeletal muscle cells in vitro. *Dev.Biol.* 117:628-635
103. Foster RF, Thompson JM, and Kaufman SJ (1987) A laminin substrate promotes myogenesis in rat skeletal muscle cultures: analysis of replication and development using antidesmin and anti-BrdUrd monoclonal antibodies. *Dev.Biol.* 122:11-20
104. von der Mark K, and Öcalan M (1989) Antagonistic effects of laminin and fibronectin on the expression of the myogenic phenotype. *Differentiation* 40:150-157
105. Funanage VL, Smith SM, and Minnich MA (1992) Entactin promotes adhesion and long-term maintenance of cultured regenerated skeletal myotubes. *J.Cell Physiol.* 150:251-257
106. De la Haba G, Kamali HM, and Tiede DM (1975) Myogenesis of avian striated muscle in vitro: role of collagen in myofiber formation. *Proc.Natl.Acad.Sci.U.S.A.* 72:2729-2732
107. John HA, and Lawson H (1980) The effect of different collagen types used as substrata on myogenesis in tissue culture. *Cell Biol.Int.Rep.* 4:841-850
108. Rao JS, Beach RL, and Festoff BW (1985) Extracellular matrix (ECM) synthesis in muscle cell cultures: quantitative and qualitative studies during myogenesis. *Biochem.Biophys.Res.Commun.* 130:440-446
109. Kuhl U, Öcalan M, Timpl R, Mayne R, Hay E, and von der Mark K (1984) Role of muscle fibroblasts in the deposition of type-IV collagen in the basal lamina of myotubes. *Differentiation* 28:164-172
110. Sanderson RD, Fitch JM, Linsenmayer TR, and Mayne R (1986) Fibroblasts promote the formation of a continuous basal lamina during myogenesis in vitro. *J.Cell Biol.* 102:740-747
111. Lipton BH (1977) A fine-structural analysis of normal and modulated cells in myogenic cultures. *Dev.Biol.* 60:26-47
112. Lipton BH (1977) Collagen synthesis by normal and bromodeoxyuridine-modulated cells in myogenic culture. *Dev.Biol.* 61:153-165
113. Williams JT, Southerland SS, Souza J, Calcutt AF, and Cartledge RG (1999) Cells isolated from adult human skeletal muscle capable of differentiating into multiple mesodermal phenotypes. *Am.Surg.* 65:22-26
114. Miranda AF, Somer H, and Dimauro S (1979) Isoenzymes as markers of differentiation, in A Mauro (ed.), *Muscle Regeneration*. Raven Press, New York, pp. 453-473
115. Verdiere-Sahuque M, Akaaboune M, Lachkar S, Festoff BW, Jandrot-Perrus M, Garcia L, Barlovatz-Meimon G, and Hantai D (1996) Myoblast fusion promotes the appearance of active protease nexin I on human muscle cell surfaces. *Exp.Cell Res.* 222:70-76
116. Baroffio A, Aubry JP, Kaelin A, Krause RM, Hamann M, and Bader CR (1993) Purification of human muscle satellite cells by flow cytometry. *Muscle Nerve* 16:498-505
117. Walsh FS, and Ritter MA (1981) Surface antigen differentiation during human myogenesis in culture. *Nature* 289:60-64
118. Walsh FS, Dickson G, Moore SE, and Barton CH (1989) Unmasking N-CAM. *Nature* 339:516
119. Yablonka-Reuveni Z, and Nameroff M (1987) Skeletal muscle cell populations. Separation and partial characterization of fibroblast-like cells from embryonic tissue using density centrifugation. *Histochemistry* 87:27-38
120. Yablonka-Reuveni Z, Quinn LS, and Nameroff M (1987) Isolation and clonal analysis of satellite cells from chicken pectoralis muscle. *Dev.Biol.* 119:252-259
121. Bischoff R (1997) Chemotaxis of skeletal muscle satellite cells. *Dev.Dyn.* 208:505-515

122. Spector DL, Goldman RD, and Leinwand LA (1997) *Cells: a laboratory manual. Culture and biochemical analysis of cells*, Cold Spring Harbor Laboratory Press, New York
123. Hurko O, McKee L, and Zuurveld JG (1986) Transfection of human skeletal muscle cells with SV40 large T antigen gene coupled to a metallothionein promoter. *Ann.Neurol.* 20:573-582
124. Nakamigawa T, Momoi MY, Momoi T, and Yanagisawa M (1988) Generation of human myogenic cell lines by the transformation of primary culture with origin-defective SV40 DNA. *J.Neurol.Sci.* 83:305-319
125 Mouly V, Edom F, Decary S, Vicart P, Barbert JP, and Butler-Browne GS (1996) SV40 large T antigen interferes with adult myosin heavy chain expression, but not with differentiation of human satellite cells. *Exp.Cell Res.* 225:268-276
126 Simon LV, Beauchamp JR, O'Hare M, and Olsen I (1996) Establishment of long-term myogenic cultures from patients with Duchenne muscular dystrophy by retroviral transduction of a temperature- sensitive SV40 large T antigen. *Exp.Cell Res.* 224:264-271
127 Deschenes I, Chahine M, Tremblay J, Paulin D, and Puymirat J (1997) Increase in the proliferative capacity of human myoblasts by using the T antigen under the vimentin promoter control. *Muscle Nerve* 20:437-445
128 Lochmüller H., Johns T, and Shoubridge EA (1999) Expression of the E6 and E7 genes of human papillomavirus (HPV16) extends the life span of human myoblasts. *Exp.Cell Res.* 248:186-193
129 Bryan TM, and Reddel RR (1994) SV40-induced immortalization of human cells. *Crit.Rev.Oncog.* 5:331-357
130 Herzig M, Novatchkova M, and Christofori G (1999) An unexpected role for p53 in augmenting SV40 large T antigen-mediated tumorigenesis. *Biol.Chem.* 380:203-211
131 Gu W, Schneider JW, Condorelli G, Kaushal S, Mahdavi V, and Nadal-Ginard B (1993) Interaction of myogenic factors and the retinoblastoma protein mediates muscle cell commitment and differentiation. *Cell* 72:309-324
132 Thorburn AM, Walton PA, and Feramisco JR (1993) MyoD induced cell cycle arrest is associated with increased nuclear affinity of the Rb protein. *Mol.Biol.Cell* 4:705-713
133 Huibregtse JM, and Scheffner M. (1994) Mechanisms of tumor suppressor protein inactivation by the human papillomavirus E6 and E7 oncoproteins. *Semin.Virol.* 67:357-367
134 Hauschka SD (1974) Clonal analysis of vertebrate myogenesis. III. Developmental changes in the muscle-colony-forming cells of the human fetal limb. *Dev.Biol.* 37:345-368
135. Baroffio A, Hamann M, Bernheim L, Bochaton-Piallat ML, Gabbiani G, and Bader CR (1996) Identification of self-renewing myoblasts in the progeny of single human muscle satellite cells. *Differentiation* 60:47-57
136. Foxall CD, and Emery AE (1975) Changes in creatine kinase and its isoenzymes in human fetal muscle during development. *J.Neurol.Sci.* 24:483-492
137. Blau HM, Webster C, Chiu CP, Guttman S, and Chandler F (1983) Differentiation properties of pure populations of human dystrophic muscle cells. *Exp.Cell Res.* 144:495-503
138. Franklin GI, Cavanagh NP, Hughes BP, Yasin R, and Thompson EJ (1981) Creatine kinase isoenzymes in cultured human muscle cells. I. Comparison of Duchenne muscular dystrophy with other myopathic and neurogenic disease. *Clin.Chim.Acta* 115:179-189

139. Pegolo G, Askanas V, and Engel WK (1990) Expression of muscle-specific isozymes of phosphorylase and creatine kinase in human muscle fibers cultured aneurally in serum-free, hormonally/chemically enriched medium. *Int.J.Dev.Neurosci.* 8:299-308
140. Bevan S, Kullberg RW, and Heinemann SF (1977) Human myasthenic sera reduce acetylcholine sensitivity of human muscle cells in tissue culture. *Nature* 267:263-265
141. Merickel M, Gray R, Chauvin P, and Appel S (1981) Electrophysiology of human muscle in culture. *Exp.Neurol.* 72:281-293
142. Iannaccone ST, Li KX, and Sperelakis N (1987) Transmembrane electrical characteristics of cultured human skeletal muscle cells. *J.Cell Physiol.* 133:409-413
143. Liu JH, Bijlenga P, Fischer-Lougheed J, Occhiodoro T, Kaelin A, Bader CR, and Bernheim L (1998) Role of an inward rectifier K+ current and of hyperpolarization in human myoblast fusion. *J.Physiol.(Lond.)* 510:467-476
144. Crain SM, Alfei L, and Peterson ER (1970) Neuromuscular transmission in cultures of adult human and rodent skeletal muscle after innervation in vitro by fetal rodent spinal cord. *J.Neurobiol.* 1:471-489
145. Crain SM, and Peterson ER (1974) Development of neural connections in culture. *Ann.N.Y.Acad.Sci.* 228:6-34
146. Peterson ER, and Crain SM (1970) Innervation in cultures of fetal rodent skeletal muscle by organotypic explants of spinal cord from different animals. *Z.Zellforsch.Mikrosk.Anat.* 106:1-21
147. Miranda AF, Peterson ER, and Masurovsky EB (1988) Differential expression of creatine kinase and phosphoglycerate mutase isozymes during development in aneural and innervated human muscle culture. *Tissue Cell* 20:179-191
148. Kano M, and Shimada Y (1971) Innervation and acetylcholine sensitivity of skeletal muscle cells differentiated in vitro from chick embryo. *J.Cell Physiol.* 78:233-242
149. Kano M, and Shimada Y (1971) Innervation of skeletal muscle cells differentiated in vitro from chick embryo. *Brain Res.* 27:402-405
150. Shimada Y, and Kano M (1971) Formation of neuromuscular junctions in embryonic cultures. *Arch.Histol.Jpn.* 33:95-114
151. Shimada Y, Fischman DA, and Moscona AA (1969) The development of nerve-muscle junctions in monolayer cultures of embryonic spinal cord and skeletal muscle cells. *J.Cell Biol.* 43:382-387
152. Shimada Y, Fischman DA, and Moscona AA (1969) Formation of neuromuscular junctions in embryonic cell cultures. *Proc.Natl.Acad.Sci.U.S.A.* 62:715-721
153. Askanas V, Kwan H, Alvarez RB, Engel WK, Kobayashi T, Martinuzzi A, and Hawkins EF (1987) De novo neuromuscular junction formation on human muscle fibres cultured in monolayer and innervated by foetal rat spinal cord: ultrastructural and ultrastructural-cytochemical studies. *J.Neurocytol.* 16:523-537
154. Kobayashi T, Askanas V, and Engel WK (1987) Human muscle cultured in monolayer and cocultured with fetal rat spinal cord: importance of dorsal root ganglia for achieving successful functional innervation. *J.Neurosci.* 7:3131-3141
155. Martinuzzi A, Askanas V, Kobayashi T, Engel WK, and Gorsky JE (1987) Developmental expression of the muscle-specific isozyme of phosphoglycerate mutase in human muscle cultured in monolayer and innervated by fetal rat spinal cord. *Exp.Neurol.* 96:365-375
156. Martinuzzi A, Askanas V, Kobayashi T, Engel WK, and Di Mauro S (1986) Expression of muscle-gene-specific isozymes of phosphorylase and creatine kinase in innervated cultured human muscle. *J.Cell Biol.* 103:1423-1429

157. Park-Matsumoto YC, Askanas V, and Engel WK (1992) The influence of muscle contractile activity versus neural factors on morphologic properties of innervated cultured human muscle. *J.Neurocytol.* 21:329-340
158. Vita G, Askanas V, Martinuzzi A, and Engel WK (1988) Histoenzymatic profile of human muscle cultured in monolayer and innervated de novo by fetal rat spinal cord. *Muscle Nerve* 11:1-9
159. Markelonis GJ, and Oh TH (1978) A protein fraction from peripheral nerve having neurotrophic effects on skeletal muscle cells in culture. *Exp.Neurol.* 58:285-295
160. Oh TH, and Markelonis GJ (1978) Neurotrophic protein regulates muscle acetylcholinesterase in culture. *Science* 200:337-339
161. Markelonis G, and Tae HO (1979) A sciatic nerve protein has a trophic effect on development and maintenance of skeletal muscle cells in culture. *Proc.Natl.Acad.Sci.U.S.A.* 76:2470-2474
162. Arnold HH, Tannich E, and Paterson BM (1988) The promoter of the chicken cardiac myosin light chain 2 gene shows cell-specific expression in transfected primary cultures of chicken muscle. *Nucleic Acids Res.* 16:2411-2429
163. Rodier A, Marchal-Victorion S, Rochard P, Casas F, Cassar-Malek I, Rouault JP, Magaud JP, Mason DY, Wrutniak C, and Cabello G (1999) BTG1: A triiodothyronine target involved in the myogenic influence of the hormone. *Exp.Cell Res.* 249:337-348
164. Sastry SK, Lakonishok M, Wu S, Truong TQ, Huttenlocher A, Turner CE, and Horwitz AF (1999) Quantitative changes in integrin and focal adhesion signaling regulate myoblast cell cycle withdrawal. *J.Cell Biol.* 144:1295-1309
165. Dodds E, Dunckley MG, Naujoks K, Michaelis U, and Dickson G (1998) Lipofection of cultured mouse muscle cells: a direct comparison of Lipofectamine and DOSPER. *Gene Ther.* 5:542-551
166. Rando TA, and Blau HM (1994) Primary mouse myoblast purification, characterization, and transplantation for cell-mediated gene therapy. *J.Cell Biol.* 125:1275-1287
167. Trivedi RA, and Dickson G (1995) Liposome-mediated gene transfer into normal and dystrophin-deficient mouse myoblasts. *J.Neurochem.* 64:2230-2238
168. Albert N, and Tremblay JP (1992) Evaluation of various gene transfection methods into human myoblast clones. *Transplant.Proc.* 24:2784-2786
169. Dhawan J, Pan LC, Pavlath GK, Travis MA, Lanctot AM, and Blau HM (1991) systemic delivery of human growth hormone by injection of genetically engineered myoblasts. *Science* 254:1509-1512
170. Salvatori G, Ferrari G, Mezzogiorno A, Servidei S, Coletta M, Tonali P, Giavazzi R, Cossu G, and Mavilio F (1993) Retroviral vector-mediated gene transfer into human primary myogenic cells leads to expression in muscle fibers in vivo. *Hum.Gene Ther.* 4:713-723
171. Springer ML, and Blau HM (1997) High-efficiency retroviral infection of primary myoblasts. *Somat.Cell Mol.Genet.* 23:203-209
172. Huard J, Akkaraju G, Watkins SC, Pike CM, and Glorioso JC (1997) LacZ gene transfer to skeletal muscle using a replication- defective herpes simplex virus type 1 mutant vector. *Hum.Gene Ther.* 8:439-452
173. Baqué S, Montell E, Camps M, Guinovart JJ, Zorzano A, and Gómez FA (1998) Overexpression of glycogen phosphorylase increases GLUT4 expression and glucose transport in cultured skeletal human muscle. *Diabetes* 47:1185-1192
174. Askanas V, McFerrin J, Alvarez RB, Baque S, and Engel WK (1997) Beta APP gene transfer into cultured human muscle induces inclusion- body myositis aspects. *Neuroreport* 8:2155-2158

175. Nicolino MP, Puech JP, Kremer EJ, Reuser AJ, Mbebi C, Verdière SéM, Kahn A, and Poenaru L (1998) Adenovirus-mediated transfer of the acid alpha-glucosidase gene into fibroblasts, myoblasts and myotubes from patients with glycogen storage disease type II leads to high level expression of enzyme and corrects glycogen accumulation. *Hum.Mol.Genet.* 7:1695-1702
176. Powell C, Shansky J, Del Tatto M, Forman DE, Hennessey J, Sullivan K, Zielinski BA, and Vandenburgh HH (1999) Tissue-engineered human bioartificial muscles expressing a foreign recombinant protein for gene therapy. *Hum.Gene Ther.* 10:565-577
177. Roest PA, Bakker E, Fallaux FJ, Verellen-Dumoulin C, Murry CE, and den Dunnen JT (1999) New possibilities for prenatal diagnosis of muscular dystrophies: forced myogenesis with an adenoviral MyoD-vector. *Lancet* 353:727-728
178. Sancho S, Mongini T, Tanji K, Tapscott SJ, Walker WF, Weintraub H, Miller AD, and Miranda AF (1993) Analysis of dystrophin expression after activation of myogenesis in amniocytes, chorionic-villus cells, and fibroblasts. A new method for diagnosing Duchenne's muscular dystrophy. *N.Engl.J.Med.* 329:915-920
179. Roest PA, van der Tuijn AC, Ginjaar HB, Hoeben RC, Hoger-Vorst FB, Bakker E, den Dunnen JT, and van Ommen GJ (1996) Application of in vitro Myo-differentiation of non-muscle cells to enhance gene expression and facilitate analysis of muscle proteins. *Neuromuscul.Disord.* 6:195-202
180. Lattanzi L, Salvatori G, Coletta M, Sonnino C, Cusella DAM, Gioglio L, Murry CE, Kelly R, Ferrari G, Molinaro M, Crescenzi M, Mavilio F, and Cossu G (1998) High efficiency myogenic conversion of human fibroblasts by adenoviral vector-mediated MyoD gene transfer. An alternative strategy for ex vivo gene therapy of primary myopathies. *J.Clin.Invest.* 101:2119-2128
181. Yaffe D, and Saxel O (1977) Serial passaging and differentiation of myogenic cells isolated from dystrophic mouse muscle. *Nature* 270:725-727
182. Blau HM, Chiu CP, and Webster C (1983) Cytoplasmic activation of human nuclear genes in stable heterocaryons. *Cell* 32:1171-1180
183. Yaffe D (1968) Retention of differentiation potentialities during prolonged cultivation of myogenic cells. *Proc.Natl.Acad.Sci.U.S.A.* 61:477-483
184. Neville C, Rosenthal N, McGrew M, Bogdanova N, and Hauschka S (1997) Skeletal muscle cultures. *Methods Cell Biol.* 52:85-116
185. Jin P, Farmer K, Ringertz NR, and Sejersen T (1993) Proliferation and differentiation of human fetal myoblasts is regulated by PDGF-BB. *Differentiation* 54:47-54
186. Caviedes R, Liberona JL, Hidalgo J, Tascon S, Salas K, and Jaimovich E (1992) A human skeletal muscle cell line obtained from an adult donor. *Biochim.Biophys.Acta* 1134:247-255
187. Sejersen T, and Lendahl U (1993) Transient expression of the intermediate filament nestin during skeletal muscle development. *J.Cell Sci.* 106:1291-1300
188. Sjoberg G, Jiang WQ, Ringertz NR, Lendahl U, and Sejersen T (1994) Colocalization of nestin and vimentin/desmin in skeletal muscle cells demonstrated by three-dimensional fluorescence digital imaging microscopy. *Exp.Cell Res.* 214:447-458
189. Gullberg D, Sjoberg G, Velling T, and Sejersen T (1995) Analysis of fibronectin and vitronectin receptors on human fetal skeletal muscle cells upon differentiation. *Exp.Cell Res.* 220:112-123
190. Gullberg D, Velling T, Sjoberg G, and Sejersen T (1995) Up-regulation of a novel integrin alpha-chain (alpha mt) on human fetal myotubes. *Dev.Dyn.* 204:57-65
191. Yasin R, Kundu D, and Thompson EJ (1982) Preparation and characterization of cell clones from adult human dystrophic muscle. *Exp.Cell Res.* 138:419-422

192. Liberona JL, Caviedes P, Tascón S, Hidalgo J, Giglio JR, Sampaio SV, Caviedes R, and Jaimovich E (1997) Expression of ion channels during differentiation of a human skeletal muscle cell line. *J.Muscle Res.Cell Motil.* 18:587-598
193. Law P, Goodwin T, Fang Q, Deering M, Duggirala V, Larkin C, Florendo A, Quinley T, Cornett J, and Shirzad A (1994) Whole body myoblast transfer. *Transplant.Proc.* 26:3381-3383
194. Law PK, Li H, Chen M, Fang Q, and Goodwin T (1994) Myoblast injection method regulates cell distribution and fusion. *Transplant.Proc.* 26:3417-3418
195. Law PK, Goodwin TG, Fang Q, Hall TL, Quinley T, Vastagh G, Duggirala V, Larkin C, Florendo JA, Li L, Jackson T, Yoo TJ, Chase N, Neel M, Krahn T, and Holcomb R (1997) First human myoblast transfer therapy continues to show dystrophin after 6 years. *Cell Transplant.* 6:95-100
196. Thompson L (1992) Cell-transplant results under fire. *Science* 257:472-474
197. Morandi L, Bernasconi P, Gebbia M, Mora M, Crosti F, Mantegazza R, and Cornelio F (1995) Lack of mRNA and dystrophin expression in DMD patients three months after myoblast transfer. *Neuromuscul.Disord.* 5:291-295
198. Mendell JR, Kissel JT, Amato AA, King W, Signore L, Prior TW, Sahenk Z, Benson S, McAndrew PE, and Rice R (1995) Myoblast transfer in the treatment of Duchenne's muscular dystrophy. *N.Engl.J.Med.* 333:832-838
199. Miller RG, Sharma KR, Pavlath GK, Gussoni E, Mynhier M, Lanctot AM, Greco CM, Steinman L, and Blau HM (1997) Myoblast implantation in Duchenne muscular dystrophy: the San Francisco study. *Muscle Nerve* 20:469-478
200. Kakulas BA (1997) Problems and potential for gene therapy in Duchenne muscular dystrophy. *Neuromuscul.Disord.* 7:319-324
201. Ito H, Hallauer PL, Hastings KE, and Tremblay JP (1998) Prior culture with concanavalin A increases intramuscular migration of transplanted myoblast. *Muscle Nerve* 21:291-297
202. Moisset PA, Skuk D, Asselin I, Goulet M, Roy B, Karpati G, and Tremblay JP (1998) Successful transplantation of genetically corrected DMD myoblasts following ex vivo transduction with the dystrophin minigene. *Biochem.Biophys.Res.Commun.* 247:94-99
203. Baque S, Newgard CB, Gerard RD, Guinovart JJ, and Gomez-Foix AM (1994) Adenovirus-mediated delivery into myocytes of muscle glycogen phosphorylase, the enzyme deficient in patients with glycogen-storage disease type V. *Biochem.J.*304:1009-1014
204. Gros L, Riu E, Montoliu L, Ontiveros M, Lebrigand L, and Bosch F (1999) Insulin production by engineered muscle cells. *Hum.Gene Ther.* 10:1207-1217
205. Koh GY, Klug MG, Soonpaa MH, and Field LJ (1993) Differentiation and long-term survival of C2C12 myoblast grafts in heart. *J.Clin.Invest.* 92:1548-1554
206. Chiu RC, Zibaitis A, and Kao RL (1995) Cellular cardiomyoplasty: myocardial regeneration with satellite cell implantation. *Ann.Thorac.Surg.* 60:12-18
207. Zibaitis A, Greentree D, Ma F, Marelli D, Duong M, and Chiu RC (1994) Myocardial regeneration with satellite cell implantation. *Transplant.Proc.* 26:3294
208. Murry CE, Wiseman RW, Schwartz SM, and Hauschka SD (1996) Skeletal myoblast transplantation for repair of myocardial necrosis. *J.Clin.Invest.* 98:2512-2523
209. Dorfman J, Duong M, Zibaitis A, Pelletier MP, Shum-Tim D, Li C, and Chiu RC (1998) Myocardial tissue engineering with autologous myoblast implantation. *J.Thorac.Cardiovasc.Surg.* 116:744-751
210. Kessler PD, and Byrne BJ (1999) Myoblast cell grafting into heart muscle: cellular biology and potential applications. *Annu.Rev.Physiol.* 61:219-42:219-242

Chapter 6

Cardiomyocytes

Ren-Ke Li
Department of Surgery, Division of Cardiovascular Surgery, The Toronto General Hospital, CCRW 1-815, 200 Elizabeth St., Toronto, Ontario, Canada, M5G 2C4. Tel: 001-416-340-3361; Fax: 001-416-340-4596; E-mail: RenKe.Li@uhn.on.ca

1. INTRODUCTION

Cardiovascular disease remains the leading cause of death in the developed world. A plethora of models and techniques have been employed in attempts to clarify and modify the pathophysiological processes involved in the development of heart diseases such as myocardial hypertrophy, ischemia and reperfusion injury, atherosclerosis and their complications. Clinical studies have allowed correlation of various predictors with the development of cardiovascular disease and outcomes of various treatment strategies. Animal models permit greater isolation of physiological variables for experimental manipulation, but are still affected by multiple factors which may be difficult or impossible to control. Cell culture model systems have therefore been employed to isolate observable cellular events from the contamination of whole organism responses. Cultured cardiomyocytes have become a favored cell type for investigation of cardiac development and pathology, as well as facilitating screening of potential interventions.

1.1 Cardiomyocyte Development and Differentiation

A brief review of normal human cardiac embryology is necessary for a discussion of cardiomyocyte culture, because cardiomyocytes isolated from hearts of different ages will differ significantly in structure, growth and function.

After the primitive myocardium forms from splanchnic mesenchyme in the third week *in utero* [1], human cardiomyocytes successively undergo fetal (embryonic), neonatal, young and adult stages in development. Human cardiomyocytes proliferate in fetal and early neonatal hearts. The

M.R. Koller, B.O. Palsson and J.R.W. Masters (eds.), Human Cell Culture Vol V, 103–124.

proliferative ability of these cells decreases and cardiomyocyte differentiation and hypertrophy occur in the late neonatal, young and adult stages [2-5].

Fetal cardiomyocytes contain contractile myofilaments. However, these fetal myofibrils are more sparse, and not as well organized, as myofibrils in adult hearts. These characteristics, however, facilitate cardiomyocyte proliferation in fetal hearts, because for hyperplasia or cytokinesis to occur, the sarcomeres within these cardiomyocytes must first dissociate and then reassemble. Cardiomyocyte hyperplasia continues at birth but declines significantly during the early neonatal period [6-8], as cardiomyocyte differentiation increases [2]. After cytokinesis halts, cardiomyocytes continue to synthesize DNA, increasing their cellular DNA content. The uncoupling of karyokinesis from cytokinesis results in the appearance of bi- and poly-nucleated cardiomyocytes. These multinucleated cells undergo progressive hypertrophy.

Cardiomyocytes in the young and adult myocardium are differentiated, a process which has been thought to be irreversible *in vivo*. These cells lose their proliferative capacity, and are unable to reorganize their intracellular contractile proteins. Because the total number of cardiomyocytes is relatively stable after the early postnatal stage [9], cardiomyocyte hypertrophy is essentially the only means of further heart development from the late postnatal stage to adulthood.

The process of cardiomyocyte proliferation and differentiation from embryonic to adult stages is complex. Several regulatory genes and proteins such as *c-myc* [10, 11], *c-fos* [12], myotrophin [13], *h-ras* [14], fibroblast growth factor (FGF) [15], and insulin-like growth factor-I (IGF-I) [16] have been implicated. The underlying mechanisms by which gene expression is controlled, and which ultimately lead to cardiomyocyte hyperplasia, differentiation and hypertrophy, are still unknown.

Several morphological and biological studies have suggested, however, that under specific conditions, young and adult cardiomyocytes may retain their capacity to proliferate. For example, apoptosis occurs during heart development and the remaining cardiomyocytes are able to proliferate to replace those that are lost [17, 18]. A limited proliferative capacity has also been reported following myocardial infarction or injury [17, 19, 20]. In end-stage heart failure, proliferating cell nuclear antigen (PCNA) has been noted in a small percentage of cardiomyocytes, along with cell mitosis [19]. Hemodynamic overload may stimulate cardiomyocytes to re-express genes usually observed only in the embryonic stage and during cell proliferation. These studies suggest that adult ventricular cardiomyocytes *in vivo* are not terminally differentiated. Cardiomyocyte hyperplasia in the adult heart may, in fact, occur in response to specific pathophysiological conditions.

1.2 Cardiomyocyte Cultures

Cardiomyocytes have been isolated and cultured at various developmental stages from the hearts of many different species, including human [21, 22], rat [23, 24], chick [25], guinea pig [26], rabbit [27] and monkey [28]. Cultured cardiomyocytes isolated at different developmental stages have widely varying physiological, biochemical and cell biological characteristics. Cardiomyocytes isolated from fetal myocardium proliferate *in vitro*, and their growth curves and rates of DNA synthesis have been well documented [22, 27, 29, 30]. These cultured fetal cells retain their *in vivo* ability to disassemble and reassemble their sarcomeres [4]. The freshly isolated cardiomyocytes are spherical, containing contractile proteins but no sarcomeres [29]. After 24 hours in culture, the cultured fetal cells have reassembled sarcomeres, and contract regularly and spontaneously in the culture dish. These cardiomyocytes proliferate in a monolayer, and when they reach confluence, the cells connect to each other by intercalated disks and form structures resembling normal myocardium [4, 21]. The intercellular junctions permit cell-to-cell communication and synchronous contraction of all the cells in that plate [22, 30].

Cultured neonatal cardiomyocytes differ from those isolated from fetal myocardium. The proportion of cardiomyocytes isolated from neonatal hearts that retains spontaneous contractile capability decreases with increasing donor age. In most cases, the majority of the cultured cells will contain contractile proteins, but only some of the cells will beat. Those that do are histologically and morphologically similar to fetal cardiomyocytes, and can proliferate and form sarcomeres [31]. The non-beating cells also proliferate and synthesize contractile proteins [4]. These cells, however, have lost the ability to assemble these contractile proteins into sarcomeres. The disruption of normal sarcomere structure that occurs during cell isolation, and the inability of the cardiomyocyte to reorganize contractile proteins into sarcomeres *in vitro*, may account for the mechanical quiescence of most cultured cardiomyocytes.

Cardiomyocytes isolated from adult myocardium will have different characteristics that are dependent upon the cell isolation procedure. When a whole heart or large sample of myocardium is perfused with lytic enzymes, the myocardium will be digested, yielding rod-shaped cardiomyocytes. These cells can be regarded as a smaller unit of myocardium, consisting of multiple cells with intervening connective tissue. Some of these cells may possess intact intercellular junctions. The sarcomeres in most of these cells will remain well organized and therefore some of the cells will contract regularly and spontaneously in culture, while others will contract only when stimulated. Since these cells retain most of the characteristics of normal myocardium and have contractile function, they are often used for physiological studies. However, the organized sarcomeres prevent most of these cells from attaching to a culture dish, and they are therefore unable to grow *in vitro*. In long-term studies, the few cells that demonstrate attachment in culture subsequently undergo extensive morphological changes with rapid disorganization of the contractile apparatus.

In contrast, when adult myocardium undergoes complete digestion, some cells without organized sarcomeres can be isolated and cultured [21, 32]. These cells are spherical rather than rod-shaped. Fragments of sarcomeres can occasionally be seen in passage 0, but these cells undergo rapid phenotypic changes in primary culture with essentially a complete loss of sarcomeres apparent upon ultrastructural analysis. The cells attach readily to a culture dish and will then proliferate. In the passaged cells, contractile proteins such as myosin heavy chain, myosin light chain and troponin I can be identified by immunohistochemistry. The myofibrillar changes noted in these passaged, partially differentiated human cardiomyocytes are similar to those in unpassaged primary adult rat and monkey cardiomyocytes [24, 28, 33, 34].

In addition to these primary cultures of cardiomyocytes, several cell lines have also been established, primarily by the targeted expression of transforming oncogenes such as the SV40 large T-antigen oncogene [35]. The AT-1 cell line, derived from the myocardium of transgenic mice, contains the alpha-cardiac myosin heavy chain promoter driving the SV40 early region. These cells have some characteristics of myocardium, including contractile proteins and sarcomeres. Because AT-1 cardiomyocytes were originally isolated from a transplantable cardiac tumor, these cells can proliferate in culture and display spontaneous contractile activity [36]. Neoplastic cardiomyocytes have also been developed by inducing expression of the *v-myc* oncogene [37]. Engelman et al. [38] transfected fetal rat cardiomyocytes with an oncogene in a retroviral vector in order to establish a cell line. These cells contained sarcomeres and contracted spontaneously. In addition to these tumor-like cardiomyocyte cell lines, embryonic stem cell-derived cardiomyocytes have also been studied. Embryonic stem cells can, under some conditions, express cardiac alpha- and beta-myosin heavy chain [39], tropomyosin [40], and myosin light chain [41], and take on many of the characteristics of cardiac myocytes [42]. They can contract, demonstrate withdrawal from the cell cycle, and become multi-nucleated when maintained in culture [43].

2. TISSUE PROCUREMENT AND PROCESSING

2.1 Human Myocardial Tissue

Although some human cardiomyocyte cell lines can be obtained from cell banks, such as the European Collection of Animal Cell Cultures (Salisbury, Wiltshire, U.K.), these cell lines have lost some of the characteristics of normal cardiomyocytes, and the potential introduction of foreign genes may be a concern. Many investigators therefore still prefer to use primary cultures obtained from myocardial tissue. However, for particular studies, the homogeneity of these banked cell lines may provide an advantage.

Obtaining fresh samples of human myocardium from which cardiomyocytes can be isolated is difficult. Potential sources of myocardium include tissue obtained from termination of pregnancy, donor hearts which are judged to be unsuitable for transplantation, and tissue removed during corrective cardiac surgery. The use of human tissue for research purposes poses a number of ethical concerns, particularly when fetal tissue is being considered. A careful review and approval of the research protocol from the institution's ethics review committee and adherence to the guidelines of this committee are crucial, as is meticulous attention to obtaining fully informed consent from the subjects.

The cells isolated from fetal myocardium or transplant donors can be considered to be healthy cardiomyocytes. However, the tissue obtained from patients undergoing corrective or palliative cardiac surgical procedures, as is more common, may be normal or abnormal, possibly reflecting either regional hypertrophy or ischemia, or a global cardiomyopathy or myocarditis. The origin of this tissue may therefore significantly affect the physiology and metabolism of these cells, and therefore the results obtained from cell culture studies. Further, the quantity of myocardium which can be obtained is also dependent on tissue origin. Some forethought directed to the potential effect of these variables on the outcomes of the studies planned is required. In addition, careful documentation of the patient's age, sex, diagnosis and location of the biopsy are crucial, but patient confidentiality must be preserved.

2.2 Cell Isolation

Cardiomyocytes are isolated from myocardial biopsies by a protease digestion technique, employing either perfusion with, or immersion in, a solution containing a combination of collagenase, trypsin, elastase and/or DNase.

2.2.1 Obtaining spherical cells from fetal, pediatric and adult myocardium

2.1.1.1 Fetal heart cells

Because fetal myocardium is not well developed and has less extracellular matrix, enzymatic digestion alone is generally sufficient to separate the cardiomyocytes. Following digestion, the isolated cells are spherical in appearance. Since they do not contain organized sarcomeres, these cells will attach readily to a culture dish and proliferate *in vitro*.

The heart and great vessels are quickly excised and placed in ice-cold Hanks' balanced salt solution (HBSS), without calcium or magnesium (GIBCO BRL), containing 100 U/mL heparin, prior to transport to a cell culture room. The residual intracardiac blood is washed out with HBSS and the aorta is then cannulated. The heart is perfused at 37°C for 15 minutes in a Langendorff apparatus with Tyrode's solution (117 mM NaCl, 11 mM glucose, 4.4 mM $NaHCO_3$, 5.7 mM KCl, 1.5 mM KH_2PO_4, 1.7 mM $MgCl_2$, HEPES 20 mM, adjusted to pH 7.4). After the 15 minute period of perfusion to wash out blood, the heart is then perfused with 1.5 mg/mL collagenase (type II, Worthington Biochemical Corp.) in calcium-free Tyrode's solution filtered through a 0.8 μm Millipore filter. The enzyme solution is maintained at 37°C in a water bath, and gassed with 100% O_2. The heart is perfused continuously with the recirculated, filtered enzyme solution for 15-20 minutes, during which the heart becomes noticeably soft and mushy, and is then removed from the perfusion apparatus. After the atria and great vessels are removed, the ventricular myocardium is transferred into cell culture dishes and minced into small pieces with scissors. The tissue fragments are then placed into 10 mL of Hanks' solution containing 1% collagenase and 0.25% trypsin, and agitated in a 37°C water bath for 5-15 minutes. The large tissue fragments are allowed to settle and the supernatant is then transferred to another tube, where an equal volume of culture medium is added. The settled tissue fragments undergo another cycle of digestion with 10 mL of collagenase and trypsin for 5 minutes at 37°C with gentle agitation. After each of 2-4 cycles of digestion, the supernatant is added to previous supernatant samples, and the remaining tissue fragments undergo repeat digestion. This pooled cell suspension is then poured through a cell strainer (Fisher Scientific) and the filtered suspension centrifuged at 580*g* for 3 minutes. The supernatant is removed and the cell pellet is washed and centrifuged two more times with HBSS to wash out the proteases. The final cell pellet is re-suspended in Iscove's modified Dulbecco's medium (IMDM, Canada Life Technologies Inc., Burlington, Ontario) containing 10% fetal bovine serum (FBS), 0.1 mmol/L β-mercaptoethanol, 100 U/mL penicillin, and 100 mg/mL streptomycin.

If the heart cannot be cannulated and perfused with the enzyme solution, an alternative is the direct digestion of the fetal tissue with collagenase and trypsin. With this technique, the myocardium is washed with HBSS and then minced with sterile scissors into 1-2 mm fragments. The myocardial fragments are digested in 10mL of HBSS containing 1% collagenase and 0.25% trypsin for 15 minutes in a 37°C water bath with gentle agitation. The supernatant is transferred to a tube containing 30 ml of culture medium. Another 10 mL of enzyme solution is added to the tissue and the digestion repeated for another 10 minutes. The supernatant is again transferred and combined, and the digestion of the tissue fragments is repeated. After 3-4 cycles, the combined cell suspension is filtered, centrifuged, washed and re-suspended in culture medium as described above.

2.2.1.2 Pediatric and adult heart cells

HBSS is removed by aspiration and the biopsies are washed three times with phosphate-buffered saline (PBS) (NaCl, 136.9 mM; KCl, 2.7 mM; Na_2HPO_4, 8.1 mM; KH_2PO_4, 1.5 mM; pH 7.3). The biopsies are transferred to a 35 mm culture dish where the white fibrous tissue is removed by careful dissection. The remaining myocardium is minced finely in 1 mL PBS prior to being transferred with the PBS to a 50 mL tube. The minced tissue is allowed to settle for approximately 1 minute. The PBS is then carefully removed and the tissue digested for 15 minutes at 37°C in 4 mL of an enzyme solution containing 0.2% trypsin and 0.1% collagenase in PBS. The supernatant is transferred to a tube containing 30 mL of culture medium. Another 2 cycles of enzymatic digestion are carried out, and the combined supernatant is centrifuged for 5 minutes at 580*g*. After discarding the supernatant, the cell pellet is resuspended in 10 mL of culture medium [32].

2.2.2 Obtaining rod-shaped cells from adult myocardium

Human atrial or ventricular myocardium can be obtained from patients undergoing coronary bypass surgery, valve repair or replacement, closure of interatrial or interventricular septal defects or from explanted recipient hearts or donor hearts which are not suitable for transplantation. Protocol review, approval and informed consent are obtained as described above, prior to obtaining these tissues.

After the tissue is removed from the heart, it is immersed in ice-cold HBSS or cardioplegic solution. Because these tissue samples are obtained in the operating room, cardioplegic solution is generally recommended because it is accepted by both the surgeon and the scrub nurse. The biopsies are then transported to the cell culture facility. Attention to maintaining the sterility of the biopsies is mandatory, with further steps in the cell isolation process carried out in a laminar flow hood. The tissue is washed with HBSS in the culture dish to remove remnants of blood and the cardioplegic solution.

When a large piece of myocardium is obtained, a careful search for macroscopically visible blood vessels is performed. If a large blood vessel is found, it may be possible to cannulate this vessel and perfuse the sample with a protease solution. The specimen is first perfused with 20 mL of HBSS to remove any residual blood, followed by 20 mL of HBSS containing 1% collagenase and 0.25% trypsin. Perfusion with protease solution is repeated until the tissue changes color from red to yellow, after which the specimen is incubated at 37°C for 30 minutes. After incubation, the tissue becomes mushy and can be minced finely with scissors. The minced tissue fragments are again incubated in the protease solution for 5-10 minutes with gentle shaking. The digested tissue is filtered and the isolated cells are washed and cultured as described in the fetal cell isolation section.

If there is no large blood vessel in the myocardial tissue, the enzyme solution can be injected into the specimen prior to incubation and cell isolation as described above. Alternatively, the tissue can be minced finely without injection of the protease solution, then washed with HBSS 3 times for 4 minutes each time. This process will remove blood and extracellular calcium. The tissue is then incubated with collagenase and trypsin solution for 15 minutes at 37°C. The cells are isolated by gentle repeated pipetting of the tissue. After transferring the supernatant to a tube containing 30 mL of culture medium, fresh enzyme solution is added to the remaining tissue and repeated cycles of digestion are carried out as described above until the yield is maximal. The cell suspension is then centrifuged and the pellet is re-suspended and used for culture.

3. ASSAY TECHNIQUES

3.1 Morphology

Cardiomyocytes, endothelial cells and fibroblasts comprise more than 95% of normal myocardium. Because these cells have substantially different morphologies in cell culture, cellular morphology can be used as an aid in the identification of cardiomyocytes. Identification is a simple procedure that does not require fixation of, or injury to, the cells. However, variability in morphological appearance limits the accuracy of this technique. Combining morphological identification with immunohistochemical staining for proteins specific to each cell type permits greatly increased accuracy.

Cardiomyocytes isolated from fetal myocardium are spherical (Figure 1a). After these cells attach to a culture dish and begin to proliferate, they take on a rectangular appearance (Figure 1b). They begin to beat rhythmically and spontaneously in cell culture. Myotubes are formed when the fetal cardiomyocytes become confluent (Figure 1c). The cells form links and take on an appearance reminiscent of normal myocardium. Some cells will not beat, but still contain substantial quantities of contractile proteins in their cytoplasm. They remain morphologically distinct from vascular endothelial cells and myocardial fibroblasts.

Complete digestion of pediatric and adult myocardium yields cells similar to those isolated from fetal tissue. The freshly isolated cells are round, but become rectangular once in culture (Figure 2a). They differ morphologically from endothelial cells, which are smaller and oval (Figure 2b), and fibroblasts, which are spindle shaped (Figure 2c). The three cell types are readily distinguished under a light microscope [21].

In contrast, cardiomyocytes isolated from adult myocardium using a vascular perfusion technique are initially rod-shaped (Figure 3), but remain morphologically distinct from the other cell types.

3.2 Growth Characteristics

The growth characteristics of cardiomyocytes differ from those of endothelial cells and fibroblasts. Fetal cardiomyocytes proliferate in culture and maintain many of the properties of cardiomyocytes *in vivo*, but they can be rapidly overgrown in a mixed culture by myocardial fibroblasts. Cardiomyocytes isolated from pediatric and adult myocardium proliferate more slowly than fetal cells, and are therefore even more susceptible to overgrowth by fibroblast contaminants. Purification of each desired cell type is therefore required in order to prevent overgrowth by rapidly proliferating cell types.

Rod-shaped cardiomyocytes are actually aggregates of cells with organized sarcomeres and extracellular matrix. These properties limit cellular adhesion to the culture dish and inhibit proliferation *in vitro*. With time, these cells demonstrate alterations in intracellular and intercellular organization, including fragmentation of the contractile apparatus and a progressive decrease in contractile myofibrils [33, 34]. The gradual loss of sarcomere integrity facilitates cellular attachment to the culture dish, but the growth pattern still differs from that of initially spherical cardiomyocytes.

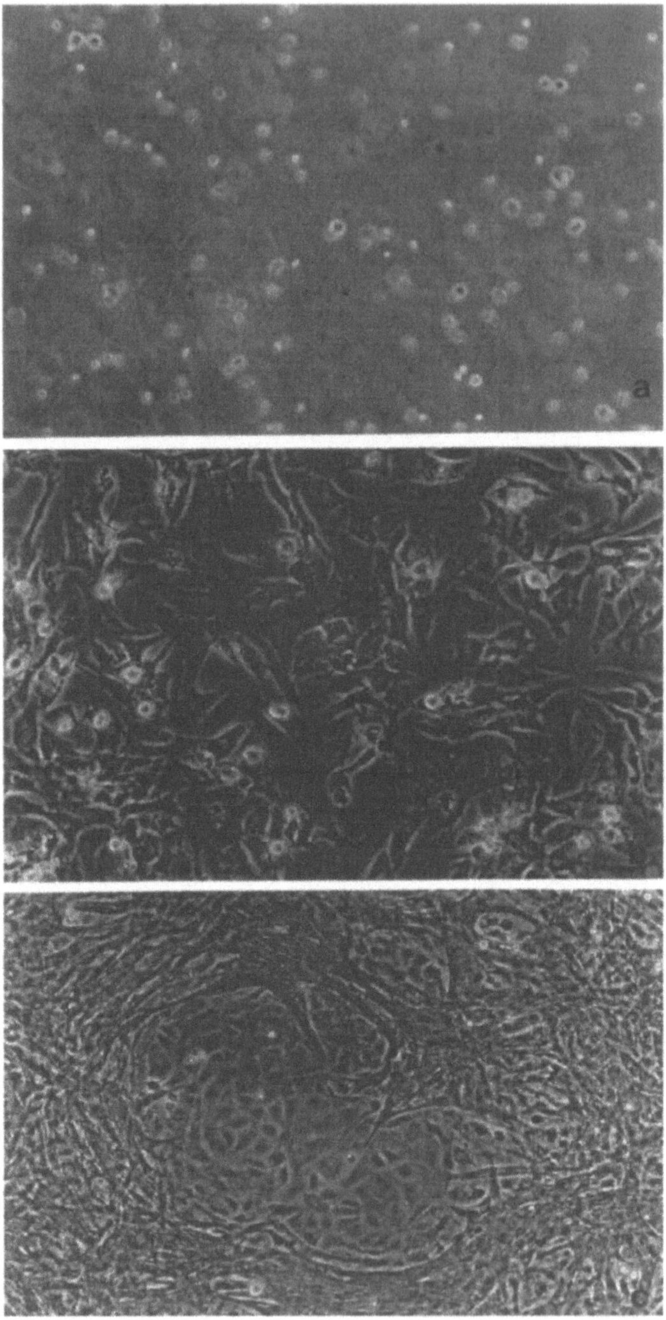

Figure 1. Freshly isolated fetal cardiomyocytes. (a) The spherical cells float in the culture medium. (b) The cultured cells are rectangular. (c) When the culture becomes confluent, the cells will link and form a tissue-like pattern (100X).

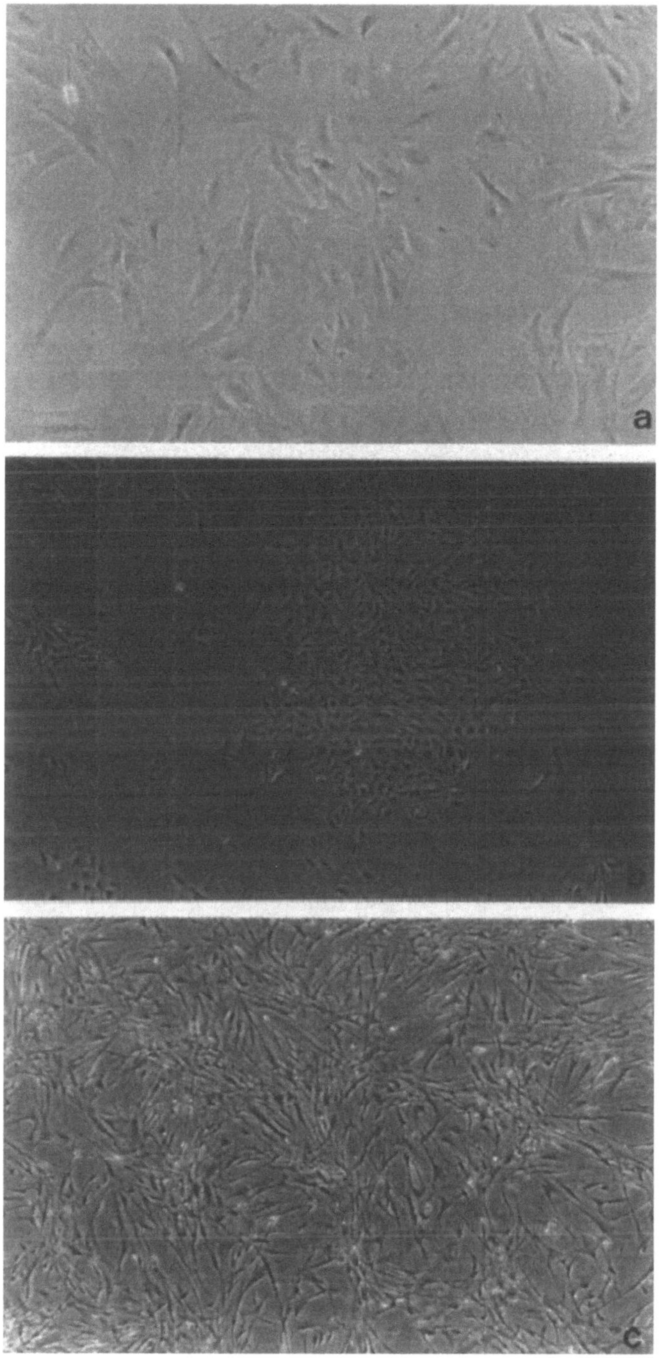

Figure 2. Light microscopic appearance of cultured human adult cardiomyocytes (a, 200X), vascular endothelial cells (b, 40X), and fibroblasts (c, 40X).

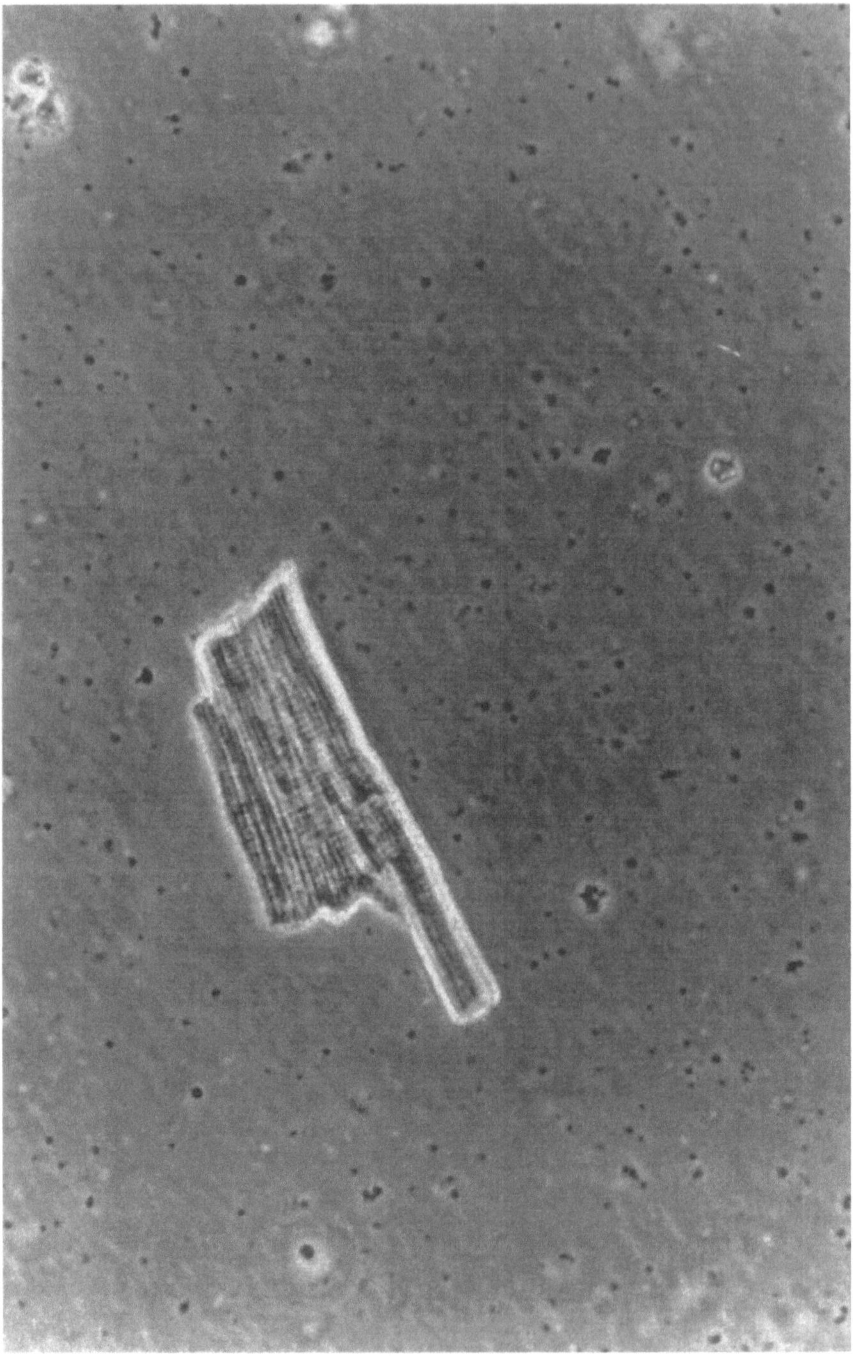

Figure 3. Light microscopic appearance of freshly isolated human rod-shaped cardiomyocytes (40X).

3.3 Myocardial Contractile Proteins

In addition to cellular contractility, the evaluation of myocardial contractile protein content within the cultured cells is a critical confirmatory step in cell identification. This can be accomplished by either Western blot analysis or immunohistochemical staining. Western blot analysis permits quantification of the levels of contractile protein in cultured cells, and therefore allows comparison between different time points, different cultures or different cell types. However, this technique yields only total protein levels in a group of cells, and does not allow quantification of the percentage of cardiomyocytes and non-myocytes in the culture.

In order to evaluate the percentage of cells in a particular culture that contain myocardial contractile proteins, immunohistochemical staining using antibodies against cardiomyocyte-specific proteins must be employed. Antibodies against the cardiac-specific troponin I isoform, human ventricular myosin heavy chain (HVMHC) and human ventricular myosin light chain (HVMLC) are commercially available. The cultured cells are fixed with cold 100% methanol at –20°C for 15 minutes, washed three times with PBS and dried, and then exposed to antibodies against cardiac-specific proteins such as CK-MB, myoglobin, troponin I, HVMHC or HVMLC for 45 minutes at 37°C. Control cells are incubated with PBS alone under the same conditions. The cells are washed three times with PBS for 15 minutes at room temperature with gentle shaking. The secondary antibody, conjugated with fluorescein isothiocyanate, is then added, and the cells are incubated with the second antibody under dark, humid conditions for 45 minutes at 37°C. After washing with PBS, the cells in the test and control groups will be visualized under an excitation light source using an epifluorescence microscope with a blue filter. The intensity at which each cell is stained yields a semi-quantitative estimate of the concentration of that contractile protein. This approach has been employed to evaluate cellular de-differentiation in culture.

3.4 Cardiac Protein Gene Expression

Characterization of cardiac-specific gene expression permits confirmation of Western blotting or immunohistochemical staining evaluation of protein levels, as well as allowing evaluation of cellular differentiation and time-dependent de-differentiation in culture. Specific gene expression can be quantified by Northern blot analysis. As the level of contractile proteins within the cultured cells decreases with time, RT-PCR may be employed in order to permit amplification of increasingly weak signals. Both of these techniques yield an aggregate estimate of gene expression in a population of cells in culture, similar to Western blot analysis of protein levels. Neither technique allows identification of specific cells, or even of the percentage of cells that are expressing a particular gene. In order to do this, *in situ* hybridization using a radiolabelled probe against a specific contractile protein mRNA must be employed.

3.5 Cellular Ultrastructure

Electron microscopy can identify cardiomyocytes by revealing the intracellular contractile apparatus. The progressive changes in intracellular structure, particularly in myotubules, can be monitored, and these data, combined with techniques evaluating cardiac-specific gene expression, permit characterization of the time course of cellular de-differentiation *in vitro*. The most significant limitation of this technique is the ability to study only a relatively small number of cells.

3.6 Biochemical Identification

Biochemical markers of myocardial tissue, including creatine kinase mass and activity, CK-MB subunit mass and activity, and myoglobin levels can be used for identification of cultured cells, as well as the evaluation of cell de-differentiation.

4. CULTURE TECHNIQUES

4.1 Cell Purification

A crucial step in the process of culturing cardiomyocytes is cell purification. Cells freshly isolated from myocardium include a mixture of cardiomyocytes, vascular endothelial cells, smooth muscle cells and myocardial fibroblasts. If the cardiomyocytes are not purified, fibroblasts will rapidly overgrow the cultures and become the dominant cell type. Purification is best carried out in the initial stages of cell culture and can be accomplished by pre-plating, dilutional cloning, chemical selection, or a combination of these techniques.

The pre-plating technique is useful for reducing contamination by fibroblasts, taking advantage of the fact that cardiomyocytes require more time to attach to a culture dish than do other cell types. The freshly isolated cells are plated on culture dishes and cultured for 2 hours in a humidified 5% CO_2 incubator at 37°C. The supernatant containing the suspended cells is then transferred into another dish for further culturing. Because many of the fibroblasts have attached to the initial plate and are therefore not transferred with the supernatant, this second culture will be relatively depleted in fibroblasts. The continuing cultures are maintained at 37°C in humidified 5% CO_2.

Another technique by which cardiomyocytes can be purified is dilutional cloning. With this technique, the isolated cells are seeded at a density of 50-100 cells per 100 mm diameter dish and cultured at 37°C in 5% CO_2. When the cells are initially seeded at low density, the isolated cells form individual colonies after approximately 1-2 weeks in culture. At this time, colonies of cardiomyocytes (identified morphologically by light microscopy) can be picked up with a sterile Pasteur pipette and transferred to new culture dishes. If any fibroblasts are adjacent to a cardiomyocyte colony of interest, the fibroblasts are injured with a sterile needle under microscopic guidance and subsequently undergo necrosis. A potential disadvantage of this clonal dilution technique, however, is that it selects the fastest growing cardiomyocytes. If these cells have somehow upregulated their growth kinetics, they may not be representative of the general population of cardiomyocytes.

It is also possible to inhibit the growth of fibroblasts in culture by chemical means. Eppenberger-Eberhardt et al. have reported that the addition of cytosine arabinoside can inhibit fibroblast proliferation in culture [44]. However, because the actions of cytosine arabinoside may not be specific only to fibroblasts, potential effects on other cell types must be considered.

4.2 Cell Culture

In cell culture systems, rod-shaped cardiomyocytes isolated from adult myocardium do not divide, and they undergo extensive morphological changes [24, 33, 34, 45]. During the first 1 to 3 days in culture, 80-90% of the cardiomyocytes become rounded and lose their cylindrical rod-like shape. Disorganization of myofibrils takes place at this stage of culture. These cells demonstrate resorption of Z-lines and rapid disorganization of the contractile apparatus.

In contrast, fetal cardiomyocytes can grow in the early stages of culture [23]. During this process, the cells disassemble and reorganize their myofibrils. Cultures of beating cardiomyocytes can be obtained. With time, these fetal cardiomyocytes undergo differentiation, and after a few generations, lose their ability to reorganize their myofibrils. At this stage, many of these cultured cells will no longer contract spontaneously. However, even the non-contracting cells still contain cardiac-specific contractile proteins and retain many of the characteristics of normal cardiomyocytes *in vivo.*

Spherical cardiomyocytes isolated from pediatric and adult myocardium are differentiated and can grow *in vitro.* Although these cells do not demonstrate spontaneous contractile activity *in vitro*, they possess many of the characteristics of normal cardiomyocytes *in vivo*. The properties include immunohistochemical staining, Northern blot analysis of cardiomyocyte-specific contractile proteins and biochemical assays of creatine kinase MB fraction activity and mass [21, 32]. These cultured cells do, however, undergo phenotypic modulation compared to their *in vivo* counterparts. In addition to the morphological changes, which include the formation of pseudopods and development of a stellate appearance, the intracellular contractile apparatus becomes progressively more disorganized and the myofibrils appear less distinct. The intracellular content of contractile proteins also decreases with increasing time in culture. The intensity of the immunofluorescent staining for contractile proteins, such as myosin heavy chain or myosin light chain, in three month old cells is significantly less than that in freshly isolated cells, and even weaker in six month old cells [21].

In any cardiomyocyte culture, the culture medium is essential for maintenance of cell morphology, growth characteristics and phenotypic modulation. For example, the concentration of serum in the culture medium determines the phenotype of the cultured cardiomyocytes. Cardiomyocytes are normally cultured in medium containing 10% FBS [46]. In this medium, fetal cardiomyocytes demonstrate little, if any, proliferation, and left unchecked myocardial fibroblasts may eventually overgrow this culture. Nag and Cheng [33] reported mitosis and cell division in adult rat cardiomyocytes during the first week of culture employing a culture medium consisting of 90% minimum essential medium and 10% horse serum, which is similar to that used by other investigators [24, 47]. The more mitogenic culture medium stimulates proliferation of the cardiomyocytes but not differentiation, and results in modulation of the normal cardiomyocyte phenotype. When fetal cardiomyocytes are cultured in a mitogen-rich medium containing 20% FBS, 1% chick embryo extract, 50ng/ml FGF and 25ng/ml multiplication-stimulating activity factor, the cardiomyocytes proliferate in culture and can undergo 30 passages or more [46]. These cells will quickly lose their contractile apparatus. However, when cardiomyocytes are cultured in lower serum concentrations (e.g. 0-5% FBS), the cell number shows no or only a small increase, but the cells maintain their phenotype and contractile properties. For example, when fetal cardiomyocytes are cultured in a mitogen-poor medium (4% horse serum), the cells cease mitosis and undergo differentiation [46].

In addition to the concentration of serum, the medium itself may also be a crucial factor. We have tested Dulbecco's-modified Eagle's medium (DMEM), medium 199, minimum essential medium, alpha-medium, and IMDM, all with 10% FBS. We found the greatest preservation of cultured cardiomyocyte morphology over time with IMDM. With the other media, cultured cardiomyocytes underwent marked phenotypic changes and became stellate or spider-shaped after several days in culture. Although the particular characteristics of IMDM that are responsible for its optimal preservation of cardiomyocyte morphology remain to be clarified, it is possible that particular components, such as the concentration of calcium or of particular vitamins, may be important.

5. APPLICATIONS OF A CARDIOMYOCYTE CULTURE SYSTEM

5.1 Cardiac Development

Studies of cardiomyocyte differentiation are an important step in understanding the development of the normal heart. A number of *in vivo* studies have identified specific morphological, biochemical and immunohistochemical changes in cardiomyocytes during heart development [25, 48, 49]. The regulation of gene expression in cardiac cells plays a fundamental role in heart development. It has not yet been possible to devise a model to evaluate gene expression in cardiomyocytes *in vivo*. Human cardiomyocyte cultures are therefore an excellent model with which to study development of the normal heart. These cell culture models allow for cell proliferation and differentiation. Cultured cells at different stages can be used to study changes in fetal gene expression [22, 50]. Using these models, expression of a number of fetal genes has been correlated with cell maturation and differentiation. In an analogous manner, the structural and biochemical changes occurring in cardiomyocytes during cell differentiation can also be studied in these cultured cells.

5.2 Cardiac Physiology

Since fetal and adult cardiomyocytes can contract in culture, physiological studies can be performed on single cells. Patch clamping can be used to record ion channel activity. Cell surface receptor expression, translocation and activity can be studied. The effect of transmembrane channel activators and blockers on cell physiology and metabolism can be evaluated. Multicellular cardiomyocyte preparations can also be used to study the effect of stretch on ion channels [51]. This simple *in vitro* model led to the discovery of stretch-activated channels in the heart. These findings may explain the occurrence of length-dependent changes in pacemaker activity and stretch-activated arrhythmias in the whole heart.

Cardiomyocyte cultures also provide a simple model with which to study cell-to-cell interaction. Eid et al. [52] demonstrated that adult cardiomyocytes undergo phenotypic changes during adaptation to primary culture. However, co-culture of cardiomyocytes with specific non-muscle cardiac cells slowed and could even reverse this process of adaptation.

5.3 Cardiac Pathology and Cellular Defenses

Isolated cardiomyocytes can be used to evaluate the effect of various pathological stimuli on the myocardium. For example, rod-shaped cardiomyocytes were isolated from the myocardium of patients with a variety of cardiac pathologies [53-55]. A whole-cell patch-clamp technique was used to measure the hyperpolarization-activated inward current, sodium current and outward potassium current, providing important information about the effects of specific cardiac diseases on transmembrane ion channels.

Cultured cardiomyocytes have also been employed to study the effect of various hypertrophic stimuli. In a cardiomyocyte culture system, individual stimuli including growth factors, hormones, cytokines, vasoactive substances and catecholamines can be administered individually without concern for the contaminating effects of whole organ or whole organism physiology. The intracellular effects of these stimuli, including alpha-skeletal and smooth muscle actin changes, and atrial natriuretic factor, can be evaluated. From these data, potential mechanisms by which myocardial hypertrophy occurs can be hypothesized and specific therapeutic interventions proposed. For example, Simpson has demonstrated, using this cell culture system, that adrenergic receptor stimulation induces cardiomyocyte hypertrophy [56].

Cardiomyocyte culture has been employed as a model to select factors which act at a cellular level to protect heart cells from injury. Using cultured human cardiomyocytes, Bowes et al. [57] demonstrated that inhibition of poly-ADP-ribose synthetase activity reduced the cell death caused by H_2O_2. To protect the myocardium from free radical injury, we tested various antioxidants and demonstrated a cell-type specific protective effect of various antioxidants, as well as a syngergistic effect of these antioxidants [58]. These *in vitro* findings were subsequently confirmed in an *in vivo* study [59]. Insulin was also identified as a important factor by which myocardial metabolism may be favorably altered during ischemia and reperfusion, leading to a clinical trial of insulin cardioplegia to improve post-operative ventricular function in patients undergoing urgent coronary artery bypass surgery [60].

6. CONCLUSIONS

Human heart cells can be isolated, purified and cultured. Depending on donor age, the isolated heart cells may vary in structure and function. Although the cells undergo phenotypic modulation in culture, they have many characteristics of normal cardiomyocytes *in vivo*. These cells can therefore be used as an *in vitro* model for cardiovascular research. The selection of fetal or adult cells is dependent on the aims of the specific study.

Cells isolated from fetal, pediatric and adult myocardium may be spherical. These spherical cells attach to a culture dish and grow *in vitro*. Fetal cells can contract in culture. This model can be used to study the development and function of the immature heart. Primary cultures of pediatric and adult cardiomyocytes are de-differentiated, but retain many characteristics of normal cardiomyocytes including contractile protein content and enzyme activity. These cells proliferate *in vitro*, but with time, the levels of contractile proteins and cardiac enzymes gradually fall.

Rod-shaped cardiomyocytes can be obtained by partial digestion of human adult myocardium. These cells are groups of cardiomyocytes linked by cardiac junctions and extracellular matrix. They contain organized sarcomeres and retain many characteristics of adult myocardial tissue. They can be regarded as miniature samples of myocardium, and can be used to evaluate physiological and pathological changes under defined conditions. These cells, however, cannot attach to a culture dish and grow *in vitro*, and are therefore unsuitable for long-term studies or studies requiring a large population of cells.

REFERENCES

1. Moore K, and Persaud TVN (1993): The cardiovascular system. In: The developing human, 5th Ed. W.B. Saunders Company, Philadelphia, pp. 304-353.
2. Clubb FJ, and Bishop SP (1984): Formation of binucleated myocardial cells in the neonatal rat: an index for growth hypertrophy. *Laboratory Investigation* 50:571-577.
3. Zak R (1973): Cell proliferation during cardiac growth. *Am J Cardiol* 31:211-235.
4. Li R-K, Mickle DAG, Weisel RD, Mohabeer MK, and Zhang J (1995): Cardiac cell transplantation. In: Mechanisms of heart failure, P Singal, I Dixon, R Beamish, N Dhalla, eds. Kluwer Academic Publishers, Norwell, Massachusetts, pp. 337-347.
5. Anversa P, Olivetti G, and Loud AV (1980): Morphometric study of early postnatal development in the left and right ventricular myocardium of the rat. I. Hypertrophy, hyperplasia and binucleation of myocytes. *Circ. Res.* 46:495-502.
6. Adler CP, and Costabel U (1975): Cell number in human heart in atrophy, hypertrophy, and under the influence of cytostatics. *Recent Advances in Studies on cardiac structure & metabolism* 6:343-355.
7. Costabel U, and Adler CP (1996): Myocardial DNA and cell number under the influence of cytostatics. II. Experimental investigations in hearts of rats. *Virchows Archiv* 32:127-138.

8. Adler CP, and Costabel U (1980): Myocardial DNA and cell number under the influence of cytostatics. I. post mortem investigations of human hearts. *Virchows Archiv* 32:109-125.
9. Arai S, and Machida A (1972): Myocardial cell in left ventricular hypertrophy. *Tohoku J Exp Med* 108:361-367.
10. Jackson T, Allard MF, Sreenan CM, Doss LK, Bishop SP, and Swain JL (1990): The c-myc proto-oncogene regulates cardiac development in transgenic mice. *Mol. Cell. Biol.* 10:3709-3716.
11. Jackson T, Allard MF, Sreenan CM, Doss LK, Bishop SP, and Swain JL (1991): Transgenic animals as a tool for studying the effect of the c-myc proto-oncogene on cardiac development. *Mol. Cell. Biochem.* 104:15-19.
12. Barka T, van der Noen H, and Shaw PA (1987): Proto-oncogene fos (c-fos) expression in the heart. *Oncogene* 1:439-443.
13. Sen S, Kundu G, Mekhail N, Castel J, Kunio C, and Healy B (1996): Myotrophin: Purification of a novel peptide from spontaneously hypertensive rat heart that influences myocardial growth. *J. Biol. Chem* 265:16635-16643.
14. Thorburn A, Thorburn J, Chen S-Y, Powers S, Shubeita HE, Feramisco JR, and Chien KR (1993): HRas dependent pathways can activate morphological and genetic markers of cardiac muscle cell hypertrophy. *J. Biol. Chem* 268:2244-2249.
15. Spirito P, Fu Y-M, Yu Z-Y, Epstein SE, and Casscells W (1991): Immunohistochemical localization of basic and acidic fibroblast growth factors in the developing rat heart. *Circulation* 84:322-332.
16. Cheng W, Reiss K, Kajstura J, Kowal K, Quaini F, and Anversa P (1995): Down-regulation of the IGF-I system parallels the attenuation in the proliferative capacity of rat ventricular myocytes during postnatal development. *Laboratory Investigation* 72:646-655.
17. Anversa P, Palackal T, Sonnenblick EH, Olivetti G, Meggs L, and Capasso JM (1990): Myocyte cell loss and myocyte cellular hyperplasia in the hypertrophied aging rat heart. *Circ. Res.* 67:871-885.
18. Kajstura J, Mansukhani M, Cheng W, Reiss K, Krajewski S, Reed JC, Quaini F, Sonnenblick EH, and Anversa P (1995): Programmed cell death and expression of the protooncogene bcl-2 in myocytes during postnatal maturation of the heart. *Exp. Cell Res.* 219:110-121.
19. Quaini F, Cigola E, Lagrasta C, Saccani G, Quaini E, Rossi C, Olivetti G, and Anversa P (1994): End-stage cardiac failure in humans is coupled with the induction of proliferating cell nuclear antigen and nuclear mitotic division in ventricular myocytes. *Circ. Res.* 75:1050-1063.
20. Liu Y, Cigola E, Cheng W, Kajstura J, Olivetti G, Hintze TH, and Anversa P (1995): Myocyte nuclear mitotic division and programmed myocyte cell death characterize the cardiac myopathy induced by rapid ventricular pacing in dogs. *Lab Invest* 73:771-787.
21. Li R-K, Mickle DAG, Weisel RD, Carson S, Omar SA, Tumiati LC, Wilson GJ, and Williams WG (1996): Human pediatric and adult ventricular cardiomyocytes in culture:assessment of phenotypic changes with passaging. *Cardiovasc. Res.* 32:362-373.
22. Kohtz DS, Dische NR, Inagami T, and Goldman B (1989): Growth and partial differentiation of presumptive human cardiac myoblasts in culture. J. Cell Biol. 108:1067-1078.
23. Li R-K, Jia Z-Q, Weisel RD, Mickle DAG, Zhang J, Mohabeer MK, Rao V, and Ivanov J (1996): Cardiomyocyte transplantation improves heart function. *Ann. Thorac. Surg.* 62:654-661.
24. Claycomb WC, and Palazzo MC (1980): Culture of the terminally differentiated adult cardiac muscle cell:A light and scanning electron microscope study. *Dev. Biol.* 80:466-482.
25. Manasek F (1970): Histogenesis of embryonic myocardium. *Am. J. Cardiol.* 25:149-168.

26. Belardinelli L, and Isenberg G (1983): Actions of adenosine and isoproterenol on isolated mammalian ventricular myocytes. *Circ. Res.* 53:287-297.
27. Horackova M (1986): Excitation coupling in isolated adult ventricular myocytes from the rat, dog, and rabbit: Effects of various inotropic interventions in the presence of ryanodine. *Can. J. Physiol. Pharmacol.* 64:1473-1483.
28. Claycomb WC, and Moses RL (1985): Culture of atrial and ventricular cardiac muscle cells from the adult squirrel monkey Saimiri Sciureus. *Exp. Cell Res.* 161:95-100.
29. Li R-K, Mickle DAG, Weisel RD, Zhang J, and Mohabeer MK (1996): In vivo survival and function of transplanted rat cardiomyocytes. *Circ. Res.* 78:283-288.
30. Goldman BI, and Wurzel J (1992): Effect of subcultivation and culture medium on differentiation of human fetal cardiac myocytes. *In Vitro Cell Dev. Biol.* 28A:109-119.
31. Atherton B, Meyer D, and Simpson D (1986): Assembly and remodeling of myofibrils and intercalated discs in cultured neonatal rat heart cells. *J. Cell. Sci.* 86:233-248.
32. Li R-K, Weisel RD, Williams WG, and Mickle DAG (1992): Method of culturing cardiomyocytes from human pediatric ventricular myocardium. *J. Tiss. Cult. Meth.* 14:93-100.
33. Nag AC, and Cheng M (1981): Adult mammalian cardiac muscle cells in culture. *Tissue Culture* 13:515-523.
34. Nag AC, Cheng M, Fleischman DA, and Zak R (1981): Long-term culture of adult mammalian cardiac myocytes: Electron microscopic and immunofluorescent analysis of myofibrillar structure. *J. Mol. Cell. Cardiol.* 15:301-317.
35. Katz EB, Steinhelper ME, Delcarpio JB, Daud AI, Claycomb WC, and Field LJ (1992): Cardiomyocyte proliferation in mice expressing alpha-cardiac myosin heavy chain-SV40 T-antigen transgenes. *Am. J. Physiol.* 262:H1867-H1876.
36. Delcarpio JB, Lanson NA, Field LJ, and Claycomb WC (1991): Morphological characterization of cardiomyocytes isolated from a transplantable cardiac tumor derived from transgenic mouse atria (AT-1 cells). *Circ. Res.* 69:1591-1600.
37. Saule S, Merigaud JP, AlMoustafa AEM, Ferre F, Rong PM, Amouyel P, Quaatannens B, Stehelin D, and Dieterlen-Lievre F (1987): Heart tumors specifically induced in young avian embryos by the v-myc oncogene. *Proc. Natl. Acad. Sci.*, USA 84:7982-7986.
38. Engelmann GL, Birchenall-Roberts M, Ruscetti F, and Samarel A (1993): Formation of fetal rat cardiac cell clones by retroviral transformation: retention of select myocyte characteristics. *J. Mol. Cell. Cardiol.* 25:197-213.
39. Sanchez A, Jones WK, Gulick J, Doetschman T, and Robbins J (1991): Myosin heavy chain gene expression in mouse embryoid bodies. An in vitro developmental study. *J. Biol. Chem* 266:22419-22426.
40. Muthuchamy M, Pajak L, Howles L, Doetschman T, and Wieczorek DF (1993): Developmental analysis of tropomyosin gene expression in embryonic stem cells and mouse embryos. *Mol. Cell. Biol.* 13:3311-3323.
41. Miller-Hance WC, LaCorbiere M, Fuller SJ, Evans SM, Lyons G, Schmidt C, Robbins J, and Chien KR (1993): In vitro chamber specification during embryonic stem cell cardiogenesis. Expression of the ventricular myosin light chain-2 gene is independent of heart tube formation. *J. Biol. Chem* 268:25244-25252.
42. Doetschman TC, Eistetter H, Katz M, Schmidt W, and Kemler R (1985): The in vitro development of blastocyst-derived embryonic stem cell line: formation of visceral yolk sac, blood islands and myocardium. *J. Embryol. Exp. Morphol.* 87:27-45.
43. Klug MG, Soonpaa MH, and Field LJ (1995): DNA synthesis and multinucleation in embryonic stem cell-derived cardiomyocytes. *Am. J. Physiol.* 269:H1913-H1921.

44. Eppenberger-Eberhardt M, Flamme I, Kurer V, and Eppenberger HM (1990): Reexpression of alpha-smooth muscle actin isoform in cultured adult rat cardiomyocytes. *Dev. Biol.* 139:269-278.(Abstract)
45. Claycomb WC, Moses RL (1985): Culture of atrial and ventricular cardiac muscle cells from the adult squirrel monkey Saimiri Sciureus. *Exp. Cell Res.* 161:95-100.
46. Kohtz DS, Dische NR, Inagami T, and Goldman B (1989): Growth and partial differentiation of presumptive human cardiac myoblasts in culture. *J. Cell Biol.* 108:1067-1078.
47. Eppenberger EM, Hauser I, Baechi T, Schaub MC, Brunner UT, Dechesne CN, and Eppenberger HM (1988): Immunocytochemical analysis of the regeneration of myofibrils in long-term cultures of adult cardiomyocytes of the rat. *Dev. Biol.* 130:1-15.
48. Sartore S, Pierobon-Bormioli S, and Shiaffino S (1978): Immunochemical evidence for myosin polymorphism in the chicken heart. *Nature* 274:82-83.
49. Price K, Littler W, and Commins P (1980): Human atrial and ventricular myosin light chain subunits in adult and during development. *Biochem. J.* 191:571-580.
50. Tseng C, Miranda E, Di Donato F, Boutjdir M, Rashbaum W, Chen EKL, and Buyon JP (1999): mRNA and protein expression of SSA/Ro and SSB/La in human fetal cardiac myocytes cultured using a novel application of the Langendorff procedure. *Pediatric Res.* 45:260-269.
51. White E (1996): Length-dependent mechanisms in single cardiac cells. *Exp. Physiol.* 81:885-897.
52. Eid H, Larson DM, Springhorn JP, Attawia MA, Nayak RC, Smith TW, and Kelly RA (1992): Role of Epicardial mesothelial cells in the modification of phenotype and function of adult rat ventricular myocytes in primary coculture. *Circ. Res.* 71:40-50.
53. Kaab S, Dixon J, Duc J, Ashen D, Nabauer M, Beuckelmann DJ, Steinbeck G, Mckinnon D, and Tomaselli GF (1998): Molecular basis of transient outward potassium current downregulation in human heart failure:a decrease in Kv4.3 mRNA correlates with a reduction in current density. *Circulation* 98:1383-1393.
54. Hoppe UC, and Beuclelmann DJ (1998): Characterization of the hyperpolarization-activated inward current in isolated human atrial myocytes. *Cardiovasc. Res.* 38:788-801.
55. Maltsev VA, Sabbah HN, Higgins RSD, Silverman N, Lesch M, Undrovinas AI (1998): Novel, ultraslow inactivating sodium current in human ventricular cardiomyocytes. *Circulation* 98:2545-2552.(Abstract)
56. Simpson P (1985): Stimulation of hypertrophy of cultured neonatal rat heart cells through an alpha-adrenergic receptor and induction of beating through an alpha- and beta-adrenergic receptor interaction. *Circ. Res.* 56:884-894.(Abstract)
57. Browes J, Piper J, and Thiemermann C (1998): Inhibitors of the activity of poly(ADP-ribose) synthetase reduce the cell death caused by hydrogen peroxide in human cardiac myoblasts. *Br. J. Pharmacol.* 124:1760-1766.
58. Li R-K, Mickle DAG, Weisel RD, Tumiati LC, and Wu TW (1989): Effect of oxygen tension on the anti-oxidant enzyme activities of tetralogy of Fallot ventricular myocytes. *J. Mol. Cell. Cardiol.* 21:567-575.
59. Teoh KH, Mickle DAG, Weisel RD, Li R-K, Tumiati LC, Coles JG, and Williams WG (1992): Effect of oxygen tension and cardiovascular operation on the myocardial antioxidant enzyme activities in patients with tetralogy of Fallot and aorta-coronary bypass. *J. Thorac. Cardiovasc. Surg.* 104:159-164.
60. Rao V, Merante F, Weisel RD, Shirai T, Ikonomidis JS, Cohen G, Tumiati LC, Shiono N, Li R-K, Mickle DAG, and Robinson BH (1998): Insulin stimulates pyruvate dehydrogenase and protects human ventricular cardiomyocytes from simulated ischemia. *J. Thorac. Cardiovasc. Surg.* 116:485-494.

Chapter 7

Dermal Fibroblasts

Jonathan Mansbridge
Advanced Tissue Sciences Inc., 10933 North Torrey Pines Road, La Jolla, CA 92037.
Tel: 001-858-713-7831; Fax: 001-858-713-7970; Jonathan.Mansbridge@advancedtissue.com

1. INTRODUCTION

This chapter deals with the culture and properties of fibroblasts. These cells form connective tissue and are widely distributed throughout vertebrates. Morphologically, fibroblasts in culture are rather uniform, broadly amoeboid, motile cells, although biochemically individual cells may differ. For the purpose of this chapter, human dermal fibroblasts will be taken as the central paradigm, fibroblasts from other vertebrate species and other tissues being treated by comparison. Fibroblast culture methods employ a range of techniques that are widely used in the culture of many cell types. Many of these are excellently described in a practical way by Freshney [1], and reference will be made to his book for detailed protocols. Discussion in this chapter will concentrate on the properties of fibroblasts and the practical implications of these properties for fibroblast culture.

1.1 Brief history of fibroblast culture

Fibroblasts are among the easiest cells to culture from vertebrates and have, therefore, been widely used in many tissue culture studies. The earliest cultures were established in the early decades of the 20th century [2], so there has now been close to 100 years experience with this cell type. Many of the earliest tissue culture techniques were based on growing cells in blood clots. In retrospect, a major reason for success with fibroblast culture is that they play a major role in wound healing, and respond to growth factors released during blood clotting as part of their normal physiological function. They are, thus, particularly adapted to culture in this type of system and the majority of subsequent tissue culture systems are serum-based for this reason. Fibroblasts are comparatively rugged [3], and can be routinely

M.R. Koller, B.O. Palsson and J.R.W. Masters (eds.), Human Cell Culture Vol V, 125–172.

cultured from essentially any organ as the most reliably available source of primary cells. In studies of mutational changes, they have advantages over lymphocytes, in that their genome undergoes no rearrangements and, indeed, they have been used to clone animals [4, 5]. As a convenient source of adherent, primary cells, they have been used for many purposes including cell physiology, cell cycle analysis, differentiation and senescence, carcinogenesis, oncogene identification, transfection studies and production of recombinant proteins.

Most work has been accomplished using fibroblast lines derived from rodents, which have very different properties from human fibroblasts. Rodent cells easily undergo changes that give rise to immortal cell lines that can be conveniently propagated without displaying the senescent phenotype characteristically shown by human cells after subculture. These lines frequently also display anchorage-independent growth, and morphological changes and tumor formation when injected *in vivo*, characteristics that are loosely described as the "transformed" phenotype. An important series of murine cell lines were developed by Todaro and Green [6] by sub-culturing primary murine embryonic cells under defined conditions of plating density and frequency of sub-culturing. They are designated by names of the form xTy (e.g. 3T3) where "x" is the number of days between subculturing and "y" is the inoculum per 5 cm Petri dish, in units of 10^5 cells. As these cultures progressed through the first 12 – 18 passages, their growth rate declined. It then increased and the resulting cultures, which showed small changes in morphology, adhesion and the ability to grow from low inoculum densities, were immortal. While these cells are frequently described as fibroblasts, they were isolated from entire embryos and their provenance is uncertain. They are, therefore, better termed 3T3 cells rather than 3T3 fibroblasts. Recent work has shown lines derived in this way to be clonal [7]. Discussion of the techniques used with transformed rodent cell lines would cover a large part of the whole field of tissue culture. This chapter will concentrate, therefore, on properties and techniques that apply to human, primary fibroblast strains.

1.2 Ease and scope of fibroblast culture

Fibroblasts can be obtained from virtually any tissue, as they are engaged in providing the collagen that allows organs to maintain structural integrity. Since they display a migratory phenotype, they are easily extracted from tissue and frequently contaminate preparations of other cell types. Since much of medium development was performed with cells related to fibroblasts, culture conditions are well optimized for their growth, and they will frequently take over cultures of other cell types unless precautions, such as selective media, are employed.

Fibroblasts that are derived from different tissues appear very similar. They are amoeboid and migratory and take up a spindle-shaped morphology. At some level, however, this similarity is illusory. Fibroblasts that generate a ligament or tendon lay down, under *in vivo* conditions, an extracellular matrix that differs greatly from that produced by fibroblasts from the skin, the liver or the vitreous humor of the eye. Thus, cells that appear morphologically very similar may possess substantially different biochemical properties, show wide divergence in gene expression, and generate diverse extracellular matrix structures. Differences between cultures of this kind have not been extensively explored, but work that has been performed supports this conclusion [8]. In this regard, it should be noted that the predominant fibroblast type isolated from the skin is thought to be derived from the papillary rather than the reticular dermis, and produces, in three-dimensional culture, fine collagen fibers as found in the more superficial layers of the dermis.

2. FIBROBLAST ISOLATION

Fibroblasts are typically isolated from normal tissue samples by one of two methods. The first method depends on the migratory properties of fibroblasts, and the second involves enzymatic digestion. Cultures immediately after isolation from tissue and before the first sub-culturing are termed "primary" cultures. After sub-culturing, they are termed passage 1, and are thereafter described by the number of sub-culturing procedures they have undergone.

If a piece of tissue is placed on a tissue culture plastic dish with a small amount of medium and incubated overnight, it will adhere. The amount of medium may then be increased over the next 3-5 days. During this time, cells migrate out of the tissue. The earliest cells to appear are tissue macrophages, which migrate away but proliferate little and do not contribute to the final culture. They are also lost upon sub-culturing with trypsin. The macrophages are followed by fibroblasts, which adhere, migrate and proliferate under these conditions. After 1 – 2 weeks, the tissue fragments may be removed and the adherent cells sub-cultured using trypsin as described below.

As an alternative technique, a larger yield may be obtained by disaggregating the tissue with collagenase. In this case, connective tissue, such as dermis, is incubated with collagenase at a concentration of about 200 U/mL at 36.5°C for up to 48 hours in growth medium. The procedure is comparatively gentle to the cells, and a crude grade of collagenase is quite satisfactory, indeed generally preferred. It is possible that proteases other than the collagenase itself may be involved in releasing the cells. The digested tissue may be mechanically disrupted and the cells removed by gentle agitation. The cells may then be collected by centrifugation and re-plated in growth medium. In addition to giving a higher yield of cells than the explant procedure, collagenase digestion probably provides a more representative sample. This is of particular importance in situations where the activity of the fibroblasts may be compromised, such as in ulcers. Trypsin has been found less valuable, as it is more likely to damage the cells on prolonged incubation and does not degrade the major connective tissue protein, collagen.

2.1 Fibroblast populations and tissue types

Fibroblast cultures are generally considered a single population, and in some respects, such as proliferation in monolayer culture, may be treated in this way. However, cloning experiments have shown that fibroblast cultures are heterogeneous in gene expression [9-11]. Thus, a fibroblast culture is better considered as a community of different cell types. A potential consequence of this observation is that clones with differing proliferation rates might be expected to take over the culture, possibly as a succession of phenotypes, as described by Bayreuther [12]. However, experiments mixing clones has suggested that the cells are capable of interacting and may maintain an equilibrium composition [13].

3. CULTURE

3.1 Monolayer culture

Fibroblasts are the prototypical adherent cell, forming cultures that proliferate, form a confluent monolayer of spindle-shaped cells that frequently form characteristic swirling patterns, and then become quiescent (Figure 1). While the formation of multi-layered cultures is generally considered a transformation characteristic, non-transformed human fibroblasts will sometimes grow over one another, particularly in the presence of ascorbate [14] or ascorbate-2-phosphate [15], in the absence or presence of transforming growth factor-β (TGF-β) [16]. The formation of a multi-layered structure under these conditions probably depends on the deposition of extracellular matrix, a differentiated property of fibroblasts, and formation of this kind of multi-layered structure should not be regarded as indicating transformation.

3.1.1. Methods

3.1.1.1 Medium

With the long history of fibroblast culture, media have been well optimized, and Dulbecco's modification of Eagle's medium (DMEM) is the standard basal medium which may be purchased from commercial suppliers. It is routinely supplemented with 10% serum, non-essential amino acids, 2 mM L-glutamine and sometimes other ingredients, usually supplied as concentrated solutions, frequently at 100 or 1000 times the final concentration. DMEM is commercially available with either high (4.5 g/L) or low (1 g/L) glucose. Fibroblasts are frequently grown using the high concentration, on the grounds that glucose is a nutrient that is depleted by the cells, and that the higher concentration allows longer intervals between re-feeding. The final concentration in medium, however, approaches that found in a poorly controlled diabetic patient, raising the possibility of non-enzymatic protein glycation. Furthermore, fibroblast growth *in vitro* is inhibited by glucose at this high concentration [17]. As in other cases, a compromise must be struck between convenience and optimal conditions. In our experience, supplementation with pyruvate provides little benefit.

Most media used for fibroblast culture are supplemented with serum, activating a characteristic range of genes [18], many of which are involved in proliferation. As discussed above, fibroblasts are particularly activated by the products of blood clotting, probably as a physiological response to wounding. In this author's experience, little benefit is to be obtained in the case of fibroblasts, however, by the use of fetal serum, and calf serum is generally quite satisfactory. There is considerable effort being applied to the development of defined and even protein-free media for fibroblasts. Such media are valuable for the isolation of soluble cell products, for providing defined culture conditions, and, in the case of fibroblast-based therapeutics, reducing the risk of disease transmission. While important, this has not proved easy, and several of the systems remain tricky to use and are not robust. When embarking on this route, it is necessary to check that a medium will work in the particular culture system proposed, as a medium that may grow monolayer fibroblasts well may not be suitable, for instance, for extracellular matrix deposition.

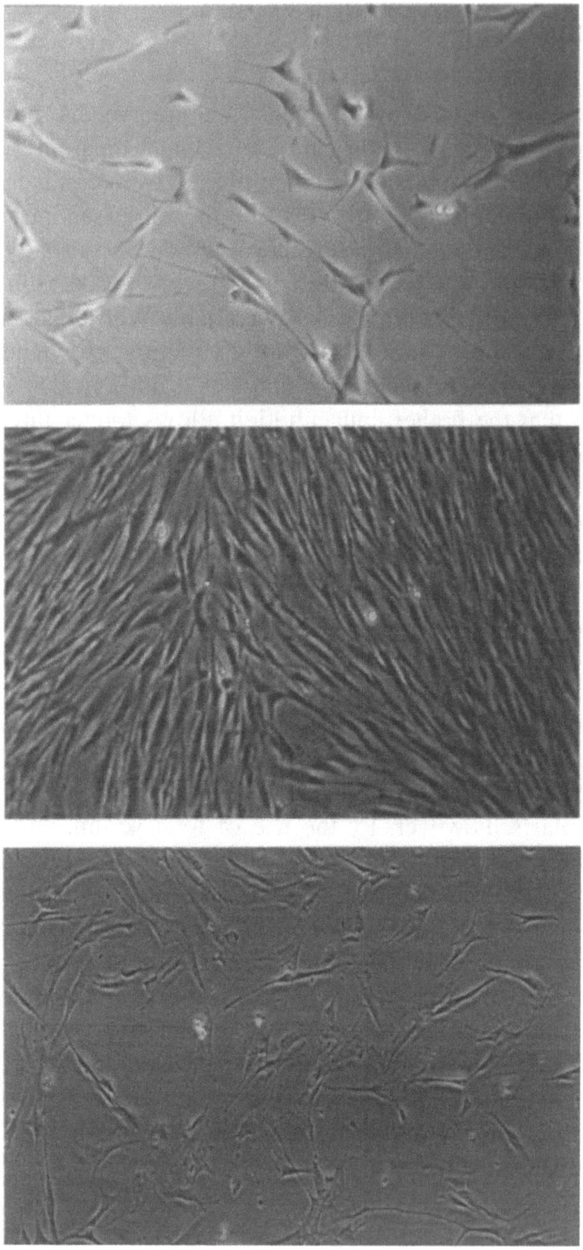

Figure 1. Fibroblasts in culture Top: sub-confluent culture at 5th passage (about 22 population doublings from isolation) in which the cells are amoeboid and migratory. Middle: (B) confluent culture at 5th passage showing characteristic swirling pattern. Bottom: culture at 12th passage (about 53 population doublings) showing senescent phenotype. Original magnification, 120 X.

The best tissue culture practice dispenses with the addition of antibiotics. The most general reason for this is that antibiotics may mask a contamination event that may not be obvious to microscopic or visual inspection, but may cause serious biochemical disturbances, rendering the cultures unsuitable for therapeutic or scientific use. One example is mycoplasma. Whether antibiotics are used or not, cultures should be routinely tested for heavy mycoplasma contamination using Hoechst 33258 staining [19] or by ribosomal RNA hybridization using one of several commercial kits (*e.g.* Gen Probe, San Diego, CA). Nested PCR kits or the use of a marker cell line can provide much more sensitive methods for detecting low level contamination.

3.1.1.2 Sub-culture

The sub-culture of fibroblasts is routinely performed using 0.05% trypsin (1:250), 0.53 mM EDTA in phosphate buffered saline (PBS), diluted from a 10X stock [20]. The designation 1:250 refers to a USP specification for trypsin activity. Medium is aspirated from the cultures, which are then rinsed with a solution such as PBS or Hank's balanced salt solution (HBSS) to remove residual serum, which contains proteins that compete with adhesion targets for tryptic activity in addition to trypsin inhibitors. The cultures are incubated with trypsin for about 5 - 10 minutes, until cells can be visibly seen free of the plastic surface, either using a microscope or by naked eye. Although there is little clear evidence, it is generally considered wise to minimize exposure of the cells to trypsin for fear of damaging cell surface receptors or the cells themselves. Some procedures have utilized digestion with trypsin on ice. The intent is to promote thorough penetration of the culture by trypsin under low temperature conditions where the enzyme is comparatively inactive, and then activate it for a short time by raising the temperature. The method has value in some cases for disaggregating thick tissue specimens, and it increases the consistency of trypsinization through the culture, although the benefit for routine sub-culturing is probably not worth the effort. The most important tryptic cleavage causing cell release is in fibronectin rather than in its cell receptor, the $\alpha_5\beta_1$ integrin. Fibroblasts are particularly easily removed from a tissue culture plastic surface, and procedures have been developed to remove fibroblasts contaminating epithelial cell cultures using EDTA alone.

As soon as the cells are released, the trypsin is quenched with trypsin inhibitor or serum. Various procedures employing serum have been used in different laboratories. The minimum is addition of an equal volume of medium containing 10% serum, but many practitioners prefer a larger amount of serum, such as 1/9 or 1/4 volume undiluted serum, or a 5–10 fold excess of 10% serum-containing medium. The cells are collected by centrifugation for 10 minutes at about 100 *g* and the cell pellet is then resuspended in growth medium by drawing into a wide bore pipette several times until all clumps have dispersed. It should not be necessary to use vigor and foaming should be strictly avoided.

Minimization of cell lysis at this stage is very important. Lysed cells release DNA, to which other cells will adhere, forming clumps that interfere with estimation of cell concentrations, and even cell seeding, potentially leading to complete loss of the cells. If the suspension has to be maintained for a protracted period before use, it should be kept cold and stirred sufficiently to keep the cells in suspension. Addition of 5 mM EDTA at pH 7.4 helps to minimize clumping.

For routine monolayer culture, fibroblasts are seeded at a density of about 5,000 cells per cm^2 into flasks or Petri dishes treated for tissue culture. The plates are incubated in a humidified incubator with an atmosphere supplemented with CO_2, which comprises a critical component of the bicarbonate buffer system. Although DMEM is formulated for a 10% CO_2 atmosphere, 5% CO_2 is frequently used. This presumably increases the pH of the medium by about 0.3 pH units, but appears to have little effect on culture performance.

3.1.1.3 Feeding

Monolayer fibroblast cultures should be fed at least every 4 days, more frequently if low-glucose DMEM is used. Spent medium is aspirated from the cultures and replaced with a suitable volume of fresh medium. Neglect of regular feeding can lead to severe loss of cells and substantial consequences in terms of senescence (see below).

3.1.2 Characteristics of Monolayer Cultures

3.1.2.1 Cloning and plating efficiency

Because of their low plating efficiency, it has been generally considered comparatively difficult to clone fibroblasts. Moreover, cloned primary human fibroblasts, possessing a limited lifespan, are of limited value. Isolation from tissue and cloning may take 30 doublings, leaving little remaining expansion potential. This limitation can be avoided using cultures immortalized though transfection with telomerase. However, cloning has been successfully accomplished by limiting dilution, which has provided insight into the population of cells in a fibroblast culture.

The earliest cloning of a fibroblast was accomplished by Earle [21] using glass capillary tubes. These studies demonstrated the importance of small medium volumes for growing single cells and the application of conditioned medium, techniques that remain important tools in successful cloning. Today, however, small culture containers can be most conveniently obtained by using 96-well, 384-well or Terasaki plates. Using systems of this kind, fibroblasts are diluted to a concentration of about 100 cells/mL and 10 μL aliquots, containing on average 1 cell/sample, are dispensed into small wells. On average, about 37% of the wells will receive a single cell. The wells should be monitored individually to check on the presence of a single cell and then incubated in a well-humidified incubator. As the clones grow, the volume of medium can be increased and the cells eventually transferred to larger containers. To be sure that the clones are derived from a single cell, it is advisable to re-clone the cells. At early stages in fibroblast cloning, feeding with supplemented DMEM containing 10%-50% fibroblast-conditioned medium is recommended, or the medium can be changed on a split feeding basis by replacing only 30% - 90% of the medium in each well. Falanga has reported that the recovery of fibroblast clones can be improved by incubation under moderately hypoxic conditions (2% oxygen) [9].

3.1.2.2 Growth characteristics

Fibroblast cultures grow by mitosis in a generally exponential manner, reaching an asymptotic value as described below. Under these conditions, the cells become quiescent in G_1, but are thought to progress to a state from which it takes progressively longer to induce them to re-enter a proliferative state. This process is poorly defined, but has been termed G_o. The quiescent state is quite distinct from senescence in at least two respects. First, the cells can be re-activated by sub-culture and, to some extent, by addition of fresh medium. Second, they do not express the various markers of senescence discussed below.

For some studies, such as those on DNA synthesis following treatment with DNA-damaging agents, it may be desirable to minimize replicative synthesis. One method is to allow the culture to reach confluence and then monitor it until the frequency of mitotic cells decreases to less than 1 per 3 low power fields. Mitotic cells are visible as spherical, refractile cells, sometimes with the spindle discernable under phase contrast, and sometimes in pairs that stand above the flattened monolayer. The cultures should be left for 1-2 days without change of medium after this point is reached, and the experiment performed in the same medium (*i.e.* without a medium change) to avoid stimulation arising from the addition of fresh medium.

3.1.2.3 Contact inhibition

Human fibroblast cultures, and untransformed fibroblasts from other species, characteristically show reduced migration and mitotic activity when in contact with other cells. This is termed contact inhibition. Although the phenomenon has long been known, there is little understanding of its mechanism. Work from Wieser's laboratory has implicated a mechanism involving a galactose-bearing cell surface protein, contactinhibin, interacting with a lectin-like receptor [22-24]. Recently, the possible involvement of connexin 43 [25] and the cadherin/catenin systems [26] have also been suggested. Contact inhibition causes changes in signal transduction [27], as might be expected, and it appears to involve $p16^{INK4A}$ at the level of cell cycle arrest [28], and $p27^{KIP}$ at the level of cell senescence [29].

3.1.2.4 Activation

It is worth bearing in mind that fibroblasts in culture express a phenotype that differs from their behavior *in vivo*. For instance, they proliferate and secrete extracellular matrix. These two properties are probably representative of a large number of changes in gene expression between a quiescent, differentiated phenotype concerned with the maintenance of mature connective tissue, and one that is concerned with wound healing. The tissue culture environment resembles wound healing not only from the dispersion of the cells, but also in the presence of wound healing growth factors derived from serum. It is generally true that fibroblasts in culture are activated to a tissue repair phenotype.

The expression array technique has recently been applied to examining the effect of serum on gene expression in quiescent fibroblast cultures [18], revealing the complexity of time-dependent gene activation and repression that occurs in a process of this kind.

3.1.2.5 Differentiation

The major differentiated function of fibroblasts, collagen deposition, is well demonstrated in three-dimensional culture. However, the cells do show other types of differentiation, notably expression of α-smooth muscle actin to form the myofibroblast. This cell type is prominent in granulation tissue. Induction of the phenotype is dependent on both TGF-β and adhesion of the cells to fibronectin containing the EIIIA splice variant [30], the type of fibronectin secreted by fibroblasts in culture and in wounds [31]. Expression of the myofibroblast phenotype is repressed by a type I collagen substratum [32] and γ-interferon. The ability to induce this phenotype appears to be a property of the majority of fibroblasts and is not restricted to a specialized fibroblast type [33], although only a sub-population may express it at any particular time.

3.2 Fibroblast Culture Lifespan

A characteristic property of human fibroblasts, originally described in detail by Hayflick [34], is their limited lifespan. In general, some 40 – 70 population doublings after isolation, fibroblasts show morphological changes, becoming large and spread (Figure 1C), and the cultures cease to proliferate (Figure 2). The phenomenon is termed cellular senescence and has been the subject of much investigation and speculation, although the mechanisms that underlie it remain poorly understood. Early hypotheses ascribed senescence to error catastrophe [35] or a stem cell/committed cell concept [36, 37]. Neither of these theories fully explains the phenomenon and each has required increasing modification [38-40]. Bayreuther noted that senescing fibroblasts pass through a well-defined series of morphotypes and show characteristic changes in proteome expression [12, 41, 42]. Detailed examination of specific gene products has shown well-defined changes [43-50], and the current view is that there is a genetically-controlled process involved [51]. By generating hybrid cells through fusing transformed cells with human fibroblasts, the senescent phenotype has been shown to be dominant over the immortality expressed by transformed and malignant cells lines [52, 53], although hybrids between early and late passage cells show intermediate phenotypes [54]. These results indicate that senescence is caused by gene expression rather than loss of function, and candidate genes have been identified on several chromosomes [55-57]. Many of the changes involve a decline in sensitivity to mitogenic stimuli that appear to involve changes in signal transduction pathways, recently comprehensively reviewed by Cristofalo [58]. Two suppressors of cell cycling, $p16^{INK4A}$ and $p21^{CIP1/WAF1}$, which inhibit cyclin-dependent kinases, have been shown to be over-expressed by cells entering senescence and by senescent cells [49, 59, 60], and are thought to account for the decline in cell proliferation. Disruption of these genes bypasses senescence in human diploid fibroblasts.

Following a different line of inquiry, changes in the terminal sequences of chromosomes, termed telomeres, have been demonstrated in aging fibroblast cultures. Chromosome telomeres have been found to carry repetitive sequences that are required for replication. DNA polymerase requires a primer on which to elongate a newly replicated strand, which, in the case of the lagging strand (the strand replicated in the direction opposite to fork movement) and in initiation of a new strand from the end of a chromosome, is RNA. This is supplied by an RNA polymerase so that the first few residues of the new chromosome are constructed from ribonucleotides, and this region is then degraded. As a result, chromosomes become progressively shorter during successive cell divisions. The terminal sequences of chromosomes consist of repeats of $d(TTAGGG)_n$ [61]. In germline [62], embryonic cells [63, 64] and possibly some cells that divide many times [65], these sequences are maintained by telomerase. In fibroblasts, however, they are generally not maintained and the length of telomeres declines with repeated cell replication [66-68] until a point is reached at which the chromosomes become unstable [69] or genes are lost.

Transfection of fibroblasts with telomerase, which regenerates the telomeric sequences, immortalizes fibroblasts [70] without causing a cancer phenotype [71]. The relationship between telomere loss and the changes in gene expression, particularly $p16^{INK4A}$, in fibroblasts is not yet clear, although separate limitation of culture lifespan by telomerase and by mechanisms related to $p16^{INK4A}$ can be clearly distinguished in other cell types [72]. A related protein that inhibits cell division, $p21^{CIP1/WAF1}$ has been found to accumulate in senescent fibroblasts [73], but is not required for the senescent phenotype [74].

In other experiments, Wilson and Jones [75] indicated that the DNA of fibroblast cultures was progressively demethylated during culture, but that methylation was restored if the cells were transformed. This suggested yet another mechanism for a senescence clock [76]. It is notable that in fibroblast lines immortalized by loss of p53, in which DNA is highly methylated, a senescent phenotype may be restored by demethylation using azacytidine [73].

The relationship between telomere maintenance, DNA methylation and expression of $p16^{INK4A}$ and $p21^{CIP1/WAF1}$ in cellular senescence has been explored by Noble *et al.* [77] and by Vogt *et al* [73]. Their results indicate that telomere length and DNA methylation operate independently [73]. Loss of $p16^{INK4A}$ will increase life span, but does not cause immortalization without another event such as telomere maintenance [77]. $p21^{CIP1/WAF1}$ appears as cultures reach senescence, while $p16^{INK4A}$ is induced later, when senescence is established. However, it is $p16^{INK4A}$ that binds to the cell cycle kinases CDK4 and CDK6 and inhibits them. $p16^{INK4A}$ can be induced by a demethylation-dependent mechanism, while $p21^{CIP1/WAF1}$ is induced by a demethylation-independent pathway.

The senescence phenomenon observed in human fibroblasts occurs both in other types of cells [78] and other animals [40, 46]. It is not easily observed in rodents, although it is possible [79], as the cultures generally develop focal colonies of cells that possess a transformed phenotype [6]. These cells are immortal, show reduced inhibition of proliferation on contact and eventually dominate the culture.

Various factors have been claimed to influence the rate of aging of fibroblast cultures. Low dose irradiation and dexamethasone have been found to extend the lifespan [80-83], whereas 5-azacytidine, which demethylates DNA, reduces lifespan [76, 84]. Cryopreservation, performed as described below, appears to have little effect on culture aging. Experiments using human fibroblasts found no difference in passage 8 fibroblasts (10:1 split) whether they had been cryopreserved at passages 6 and 7 or not. In contrast, failure to feed cultures, leading to a sharp decrease

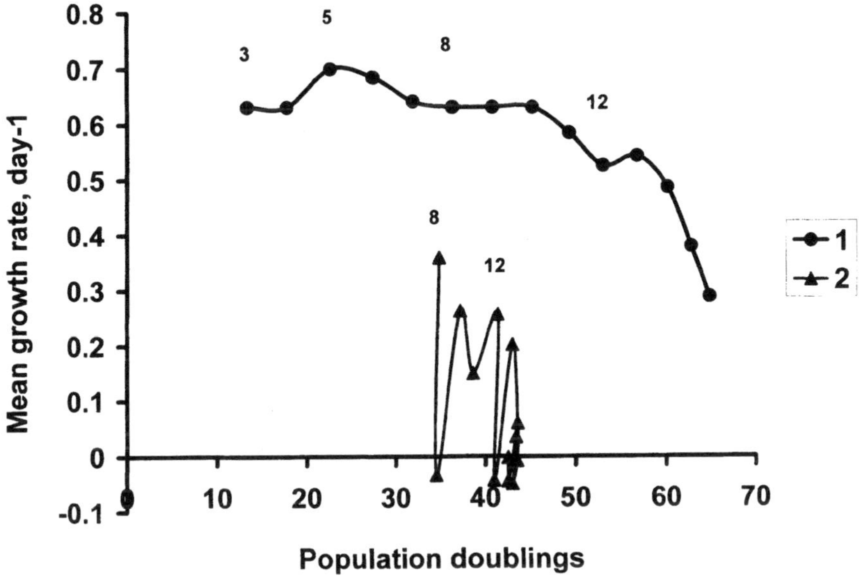

Figure 2. Changes in growth rate with population doubling. Fibroblasts were cultured as monolayers and sub-cultured weekly at a plating density of 5,000 cells per cm^2 and fed at 4 days. The cells were counted at harvest and a nominal growth rate calculated. This is plotted against the number of cumulative population doublings. The numbers above the lines are passage numbers. Senescence is visible in line 1 from about 60 doublings. In the cultures of the same cell strain used for line 2, midweek feedings were omitted at the 9th and 13th passages, resulting in major loss of viability and cells. It may be noted that the number of cells harvested from these two cultures was less than the number seeded, resulting in retrograde movement on the abscissa and a negative growth rate. Although senescence occurred at about the same passage number, the number of population doublings was reduced by about 20, although the expected number of cell doublings required to make up the cells lost should have been about 7. The conclusion from this observation is that mishandling fibroblast cultures may result in a much larger consumption of potential cell doublings than might have been expected from analysis of the event itself.

in the number of adherent cells, has been found to cost about 10 population doublings (Figure 2).

Examination of gene expression in senescent cultures using array technology has indicated that the cells activate several genes associated with inflammation [85, 86] and extracellular matrix degradation. This is consistent with observations of elevated and disregulated metalloprotease secretion [87-90], obtained by classical techniques. The observations are interesting as they may be related to cellular abnormalities found in chronic wounds.

3.2.1 Stem cell question

The existence of stem cells within fibroblast populations has been invoked to explain senescence, but the question has not been settled. No stem cell population has ever been isolated from fibroblast populations and the evidence for such cells is indirect.

3.2.2 Passage and doubling

The life of a culture is commonly measured by two different scales: passage and doubling. A cell or population doubling constitutes an average of one cell division per cell, and is the way in which fibroblast lifespans were originally expressed, giving numbers of 50 – 70. Passage number refers to the number of sub-culturing operations. Depending on how this operation is performed, it may correspond to any number of doublings, but is commonly about three cell divisions, corresponding to an 8:1 split. In this case, senescence occurs at about passage 13 - 20.

3.2.3 Practical Considerations of Fibroblast Senescence

From the practical point of view in fibroblast culture, the mechanism of senescence is irrelevant. It is important, however, to bear in mind that the cultures do change in phenotype with passage (Figure 1), and it is important to always use cultures at a comparable level of maturity.

3.2.4 Methods to maintain consistent cultures

For rigorous work, it is important to use fibroblasts at a consistent stage of their lifespan. The precise stage is a compromise between cell yield and cell quality. While the proliferation rates of fibroblasts decline quite suddenly as they become senescent (Figure 2), there is a steady decline at earlier stages. In general, this is not a limiting factor. It may be borne in mind that the yield of cells at 40 doublings, close to the practical limit, is about 10^{12} times the original inoculum. In general, cell banks are cryopreserved at various stages of the lifespan to provide convenient expansion for experiment or manufacturing. Typically, a single isolate can be frozen at passage 3, using 10:1 splits, as about 200 vials. The cells can then be thawed, grown and frozen again at a later passage.

3.2.4.1 Cryopreservation

Fibroblasts in suspension are easily and reliably stored by cryopreservation at liquid nitrogen temperatures in a cryoprotectant solution containing 10% DMSO in growth medium. Other cryoprotectants, such as glycerol, have been employed, but for suspended fibroblasts, DMSO is satisfactory. Cultures are trypsinized as described above. DMSO is toxic to some cell types, including fibroblasts in suspension, although not in three-dimensional culture [91], so it is important to minimize the time that the cells are in contact with it. Depending on the number of vials to be prepared, the cells may be re-suspended in cryoprotectant at a density of about 1.5 x 10^6 per mL, pipetted in 2 mL aliquots into freezing vials and frozen, or re-suspended in growth medium, aliquoted and then DMSO added to a final concentration of 10%. The second method has the advantage of minimizing osmotic shock to the cells, but often results in precipitation of proteins out of the serum. The cells must be cooled slowly, ideally using a controlled rate freezer set to a –0.4 to –1°C per minute cooling rate. Alternatively, it has generally been found adequate to store the vials in an insulated container, such as an expanded polystyrene foam box with a wall thickness greater than 15 mm, in a –70°C freezer for 6 hours to overnight. The vials are then transferred to liquid nitrogen storage, either in the liquid or vapor phase. The important point here is that, for long term stability, storage must be below the glass transition temperature of the water/10% DMSO system, about –120°C, which can be achieved in the vapor phase of liquid nitrogen.

3.2.4.2 Fibroblast immortalization

Fibroblast immortalization is best divided into 2 groups: human fibroblasts forming one group and those of other species forming another. Most work has been performed on rodent fibroblasts that spontaneously undergo a process that generates immortal cell lines (i.e. transformation) as discussed above. Human cells do not undergo spontaneous transformation. In a careful study in 1988, McCormick [92] concluded that up to that time, there was only one unequivocal case of spontaneous transformation of human fibroblasts [93]. Honda reported isolation of spontaneously immortalized lines from a high-dose radiation survivor [94]. Even carcinogen and radiation treatments are questionable sources of transformants from human fibroblast strains [81, 92, 95].

Human fibroblasts can, however, be immortalized by transfection with SV-40 T antigen, either in the form of the virus or on a plasmid with a selectable marker [96]. Transfection produces an immediate change in the phenotype of the cells without immortalizing them. After a number of generations, the cultures decline in growth rate and proceed through a "crisis" from which immortal lines emerge. The nature of the process is not clear, but numerous changes occur, including maintenance of DNA methylation [97], chromosomal changes [98] and stabilization of telomere length [99]. The resulting cells are immortal, but their relationship to the original fibroblast strain is uncertain.

As discussed above, recent evidence has indicated that telomere length plays a critical role in fibroblast aging and that cells can be immortalized merely by providing them with the means to maintain telomere structures. Transfection with telomerase has been found to confer immortality on human fibroblasts [70]. Currently available evidence indicates that fibroblasts immortalized in this way show no other transformed properties [71], so the method appears ideal for the production of cell strains immune to senescence.

3.3 Three-dimensional culture

Fibroblasts are frequently loosely considered to be a primitive cell type lacking the obvious differentiated functions of cells such as keratinocytes. However, although it is not commonly displayed in simple monolayer culture, fibroblasts are highly specialized in collagen deposition and maintenance and have a clear differentiated phenotype that can be demonstrated *in vitro* in three-dimensional culture.

Several three-dimensional culture systems for fibroblasts have become available. In these systems, fibroblasts behave quite differently from monolayer cultures, and the systems differ from one another to a major extent. The earliest such system was developed by Bell who cast fibroblasts in collagen gels [100-102]. More recently, fibroblasts grown on woven or knitted three-dimensional scaffolds have been found to form a dermal-like tissue [103, 104]. Even in the absence of a scaffold or collagen gel, fibroblasts under suitable conditions (*e.g.* at high density in the presence of ascorbate) will produce a dermis-like three-dimensional structure that may reach 16 cell layers thick [14-16]. Each of these three-dimensional culture systems shows properties that differ from monolayer culture.

3.3.1 Collagen Gel Cultures

Three-dimensional culture systems based on collagen gels are prepared by suspending fibroblasts in a neutralized collagen solution with growth medium, and then pouring into a Petri dish [100, 102, 105]. Collagen in solution can be obtained commercially as Vitrogen (Collaborative Research, Framingham, MA) which is derived from calf skin, or can be extracted from a source such as rat tails (e.g. see [1], page 195). Collagen is soluble at low pH and is generally supplied in dilute acetic acid. The amount of normal NaOH required to neutralize the acid should be determined by titration. Two systems of this kind have been devised, relaxed gels and stressed gels.

Relaxed gels are formed by casting the fibroblast-containing collagen gel in a dish, such as a microbiological Petri dish, to which the gel cannot adhere. The resulting free-floating gel can contract without constraint. Under these conditions, fibroblasts cause the gel to contract by a factor of 20- to 30-fold by a cAMP-dependent mechanism [106]. This produces a dermis-like structure with substantial mechanical strength, in which the fibroblasts become quiescent and comparatively unresponsive to growth factors [107, 108]. A stressed gel is produced by casting the collagen gel on a tissue culture-treated dish or in a frame so that the gel is not free to contract.

In relaxed collagen gels, the majority of the cells cease to proliferate, and the amount of collagen mRNA and the rate of collagen synthesis is greatly reduced [109-113]. In a stressed gel, a majority of cells continues to proliferate, although about 25% do not, and collagen synthesis is comparable with monolayer culture and is not completely inhibited. Collagenase (MMP-1) is induced, which is thought to be mediated through the collagen receptor, $\alpha_2\beta_1$ integrin. Integrin $\alpha_2\beta_1$ is induced by interaction with fibrillar collagen by an NFκB-dependent mechanism [114], whether in monolayer or three-dimensional culture. Both relaxed and stressed collagen gel cultures provide good substrates for keratinocyte growth and have formed the basis for the development of skin equivalent systems [101, 102, 105, 115-118]. In terms of their comparatively quiescent fibroblasts, these cultures have several properties in common with non-wounded skin.

3.3.2 Scaffold-based Three-dimensional Cultures

When cultured on a three-dimensional polymer scaffold, fibroblasts proliferate in a logistic manner as a single population nearly as rapidly as they do in monolayer culture. After about 1 week, the cells start to deposit extracellular matrix and fill up voids in the scaffold structure (Figure 3). Collagen deposition (Figure 4) is associated with a shift in energy metabolism from glycolysis to oxidation (Figure 5). The rate of collagen deposition is very high, approaching 200 pg/cell/day or 2,000 – 3,000 molecules/sec/cell. By comparison with monolayers, fibroblasts in this type of culture also induce mRNA for the inflammatory cytokines IL-6, IL-8 and G-CSF, and the angiogenic factor vascular endothelial cell growth factor (VEGF) [119]. In contrast to collagen gel-based cultures, scaffold-based cultures show many properties in common with active wound healing tissue in their high rates of proliferation, extracellular matrix deposition and cytokine induction.

Scaffold-based fibroblast culture systems have been grown on several types of support. The simplest is woven nylon fabric with a thread spacing of about 140 μm. Various other scaffolds have been used based on different materials, including a degradable co-polymer of lactate and glycolate, and on different structures, including chopped strand fiber mats, porous sponges, knitted, woven and braided fabrics. To a first approximation, the nature of the material and its physical structure appear to be of secondary importance to its three-dimensionality and a cavity size within the correct range. Cells are seeded at about 10^5 per cm^2, and then fed periodically with DMEM supplemented with 10% BCS, 2 mM L-glutamine, non-essential amino acids and 50 μg/mL ascorbic acid. At early times, the intervals between feeding can be similar to monolayer cultures (3-4 days). At later times, as the cells reach higher densities and begin to secrete collagen, the frequency of medium changes should be increased. The interval between feeding depends on the type of DMEM (high or low glucose) and may be judged by estimating glucose utilization with an instrument such as the YSI glucose/lactate analyzer (YSI, Yellow Springs, CO) or, with experience, from the color of the medium.

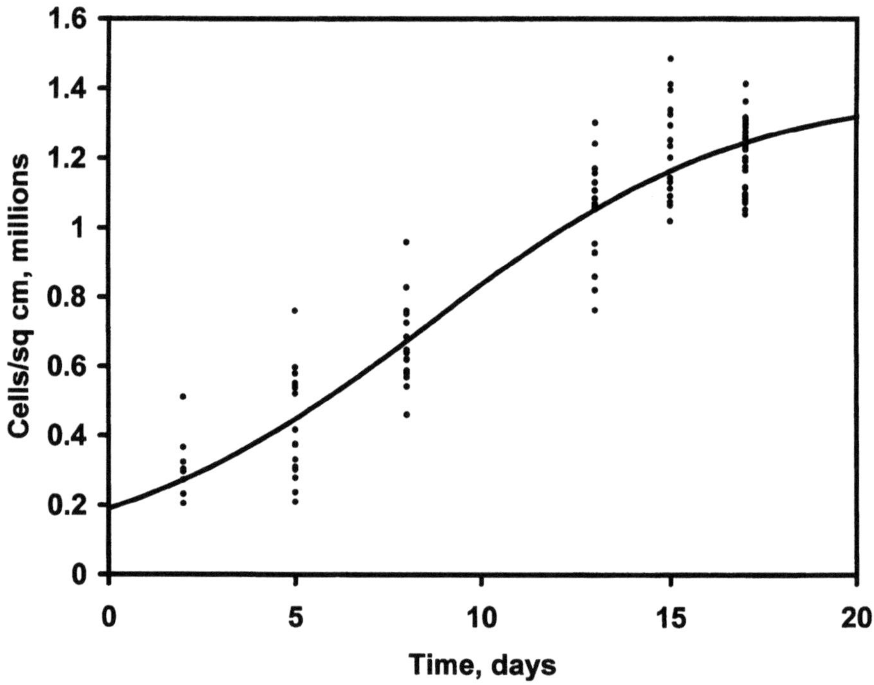

Figure 3. Proliferation data for a fibroblast culture fitted to the Verlhurst equation. Data were obtained from multiple three-dimensional cultures growing on a lactate/glycolate co-polymer scaffold in a static batch-fed system. Medium was changed when about 25% of the glucose in the medium had been consumed.

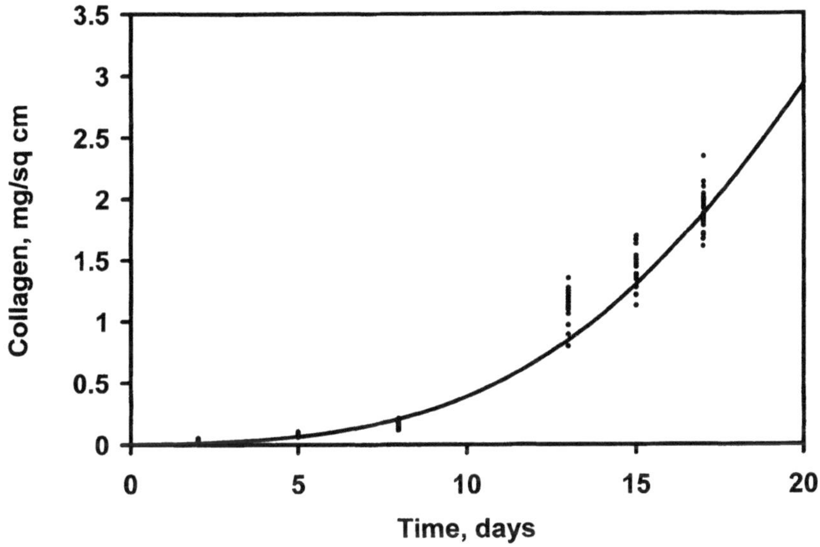

Figure 4. Collagen deposition in a three-dimensional culture. Collagen deposition was determined in the same experiments as used for Figure 3, and is fitted to the equation described in the text with the cell density raised to a power of 2.6.

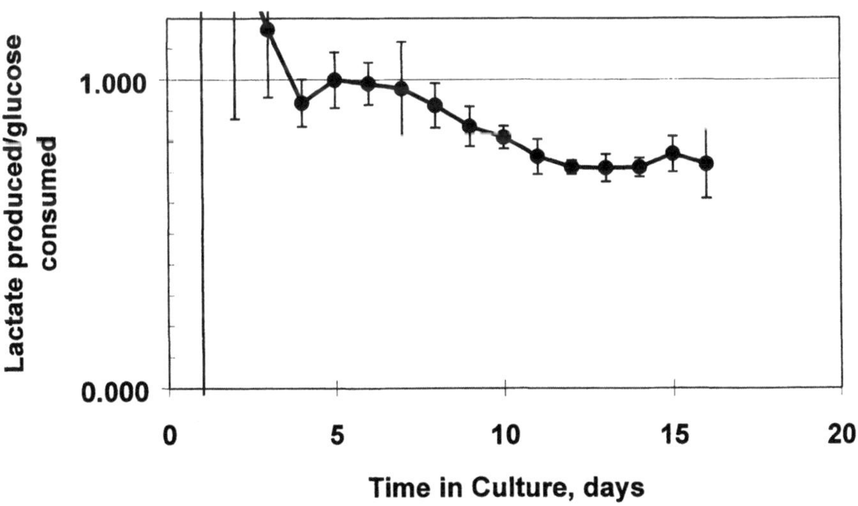

Figure 5. Changes in the ratio of lactate production to glucose consumption in three-dimensional cultures during incubation. The data are a compilation of a large number of experiments using three-dimensional fibroblast cultures grown on a nylon scaffold. Growth and matrix deposition in these experiments were comparable to those shown in Figures 3 and 4.

3.3.3 Effect of culture conditions on collagen deposition

3.3.3.1 Three dimension culture

The mechanism of control of the very high rate of collagen deposition observed in scaffold-based three-dimensional cultures is not clear. One possible explanation is related to the mechanism of collagen deposition mediated by pro-collagen C-proteinase, shown to be bone morphogenetic protein 1 (BMP-1) [120, 121]. Cleavage of the C-terminal region from pro-collagen greatly reduces its solubility in aqueous solution at neutral pH and results in precipitation of the triple helical region as fibrillar collagen. Three-dimensional culture may promote increased local concentrations of pro-collagen and BMP-1, which allows this reaction to take place in a manner that does not occur in monolayers. A similar argument may be applied to autocrine growth factors such as connective tissue growth factor (CTGF). It has also been noted that the activity of the C-terminal pro-collagen peptidase is increased 10 – 15-fold if the pro-collagen is aggregated [122]. Three-dimensional culture may well provide more favorable conditions for pro-collagen aggregation than monolayer culture.

3.3.3.2 Effect of ascorbate

The importance of ascorbic acid in the maintenance of connective tissue has been known since its discovery by Waugh and King in 1932 [123] as the curative agent found in lemons for scurvy, which was shown to be identical to a "hexuronic acid" isolated by Szent-Györgyi. Ascorbic acid appears to influence collagen synthesis in at least two ways. Firstly, it is a co-factor for proline and lysine hydroxylases, although its role generally appears to be catalytic rather than stoichiometric [124]. Since proline hydroxylation is required to stabilize the collagen triple helix to the point that it remains intact at physiological temperatures, under-hydroxylation causes intracellular collagen degradation, and hence reduced secretion [125]. This, however, does not appear to be the major effect of ascorbate on collagen synthesis and, as a second and apparently independent action, ascorbate induces collagen at the mRNA level [126-130]. As discussed above, ascorbate addition can induce the formation of a three-dimensional extracellular matrix structure.

3.3.3.3 TGF-β_1 and Connective Tissue Growth Factor (CTGF)

TGF-β_1 is the archetypal growth factor stimulating connective tissue deposition. Work from Grotendorst's group has indicated that this action is mediated, in the case of fibroblasts, through stimulating secretion of CTGF [131, 132]. The CTGF promoter carries TGF-β response elements [132] and its expression is highly induced by TGF-β_1. CTGF induces collagen synthesis and, under suitable conditions, fibroblast proliferation. The reason for this apparent growth factor relay is not yet evident, but it may be important both in providing control of extracellular matrix synthesis independent of TGF-β, and in enabling further control on one aspect of the action of TGF-β. In common with other autocrine growth factor systems [133], CTGF may also provide a chemical means of sensing the environment of the cell and controlling collagen synthesis in a manner responsive to extracellular conditions. CTGF carries several binding sites for IGF, TGF-β and heparin that may play a role in such a sensor function [134].

4. ANALYTICAL METHODS FOR THE STUDY OF FIBROBLAST CULTURES

4.1 Methods for Determining Cell Proliferation

Numerous methods are available for determining the growth characteristics of fibroblasts. Essentially any cell property that does not change with culture density can be applied. The following is a selection of the more useful methods.

4.1.1 Cell number

The simplest and most direct approach to determining the cell density of a fibroblast culture is to count the cells. This may be performed visually using a hemacytometer, or using a particle counter (e.g. Coulter, Hialeah, FL). A hemacytometer is a simple, cheap instrument that gives reliable results in experienced hands. It is, however, essential to follow the manufacturer's instructions, to take the sample from the middle of a well-mixed culture, and to fill the hemacytometer in a rapid and consistent manner. The hemacytometer is particularly useful for performing comparative counts of live and dead cells after staining with trypan blue.

Particle counters, such as the Coulter counter, once set up, provide values that are less variable from operator to operator than those obtained with a hemacytometer. It is important, in this case, that the gates are set correctly, but once this is performed, consistent results can be obtained.

4.1.2 Protein

With the possible exception of senescent cultures [135], fibroblasts retain an approximately constant protein content averaged over the cell cycle which can be used to follow culture growth. As discussed below, the value will change with the age of the culture as the cells move into stationary phase. However, the major protein component of monolayer cultures is serum that must be removed. In the case of adherent fibroblasts, this is easily accomplished by washing with a protein-free isotonic solution such as PBS. Many convenient commercial kits are available for protein estimation from suppliers such as Pierce (Rockford, MD), BioRad (Hercules, CA), and Sigma (St. Louis, MO).

4.1.3 DNA

Diploid cells contain a constant amount of DNA, 6 pg in humans [136, 137], so DNA estimation may be used to determine cell number. In a proliferating population, however, a proportion of the cells are in S and G_2 phases, so the average cellular content is somewhat higher. For fibroblasts dividing about once every 24 hours, a value of 7 pg/cell is appropriate. Clearly, this value changes with the growth phase of the culture, and to use this method as an accurate means of determining cell number requires estimating the distribution of cells in different phases of the cell cycle by flow cytometry. Commercial kits for DNA estimation methods with sensitivity suitable for this purpose are widely available, generally using the binding of fluorescent dyes such as Hoechst 33258.

4.1.4 Metabolic activity

Metabolic activity, determined by a reduction of dyes such as 3-[4,5-dimethylthiazol-2-yl]-2,5-diphenyltetrazolium bromide (MTT), 2,3-bis[2-methoxy-4-nitro-5-sulphophenyl-2H-tetrazolium-5-carboxanilide (XTT), or Alamar Blue, has been used to follow changes in culture properties both during growth and toxicity testing. The dyes are reduced to intensely colored formazan derivatives that may be insoluble (MTT) or soluble (XTT, Alamar Blue) in medium. Their concentrations may be determined spectrophotometrically either directly in the medium (XTT, Alamar Blue), or after extraction with isopropanol (MTT). These methods show high precision and ruggedness, but, as with protein and DNA contents, metabolic activities measured by these means in individual cells can vary substantially and change with the growth phase of the culture.

MTT reduction has been widely used for estimating cell viability in cytotoxicity [138-145], and for following cell proliferation [146-149] in growth factor assays. The use of MTT for this purpose stems from the work of Slater on mitochondrial electron transport [150], and it has been generally considered that, in mitochondria, MTT is reduced mainly (60%) by ubiquinone through reduction by succinate dehydrogenase, but also by cytochrome c (40%). However, MTT is efficiently reduced by NADH, and studies by Berridge indicate that much of the cellular MTT reduction is extra-mitochondrial [151]. It thus appears that the assay measures the activity of part of the cytochrome electron transport chain and part of the citric acid cycle, together with extra-mitochondrial dehydrogenase systems. Provided the amount of these activities per viable cell remains constant and activity is lost in dead cells, MTT reductase activity provides a useful measure of the number of viable cells. However, it must be noted here that cellular MTT reductase activity is not necessarily constant. It was recognized early [146] that MTT reduction could be used to detect cellular activation independently of proliferation. Berridge [152] showed that cAMP would increase MTT reduction, and Vistica [153] found that MTT reduction could be greatly influenced by the glucose concentration in the medium. However, provided the cells are compared under the same conditions of medium and cell activation, MTT has been found to be a good measure of viable cells.

Although the formazan derived from MTT is insoluble and precipitates in cells, MTT reduction by monolayer fibroblast cultures is linear for 2 hours. The relationship between dye reduction and MTT concentration shows a maximum at a concentration of about 1 mg/mL. Thus, cultures may be incubated with MTT at a concentration of 0.5-1 mg/mL in medium supplemented with 2% - 10% serum, at 37°C for a standardized time up to 2 hours. The medium is removed, the cultures washed with PBS or HBSS and the formazan extracted with isopropanol. The blue color is conveniently determined in a plate reader at 540nm. Clearly, some variation is possible in application of the method to suit different purposes, although parameters such as the time of incubation with MTT, the volume of isopropanol used to extract the formazan and the temperature of incubation need to be applied consistently.

4.1.5 Dye dilution

Monitoring cell division by determining the dilution of a membrane-binding tracker dye is an approach to estimating cell proliferation that differs fundamentally from any of the methods that estimate increases in a cellular property, and provides quite different information about the culture. This technique is based on the development of fluorescent cell tracker dyes that bind to and associate irreversibly with the cells [154-157]. This has been developed into a system for following the generational composition of a population, using the division of the cellular dye load between the two daughter cells during mitosis [158]. The cells are labeled, excess dye removed and a reference distribution of the dye in the cell population is determined by flow cytometry. The cells are then inoculated into culture and, as the cells divide and the dye is distributed between the daughter cells, the flow cytometric pattern shifts and broadens to lower fluorescence intensity.

Using the original flow cytometric distribution as a reference, the distributions obtained at later times can be resolved into a series of Gaussian curves from which the generational compositions or the populations can be calculated. This methodology allows the detection of sub-populations in the culture that proliferate at different rates. It has been developed into a commercial kit (Cell Census Plus™, Sigma). The generational compositions can be compared with predictions from the Verlhurst equation (see below), including if necessary, multiple populations. Using this method, it has been found that non-senescent fibroblast cultures grow both in monolayer and in scaffold-based three-dimensional culture as a single population. In contrast, collagen gel-based three-dimensional fibroblast cultures show multiple populations, including a proportion of non-proliferating cells [119].

4.2 Mathematical description of culture growth

The growth of a fibroblast culture is logistic, starting with a lag phase (frequently related to the time spent in stationary phase prior to sub-culture), a log phase, then leveling off to stationary phase as the cells become contact inhibited. The three values that characterize this type of growth are the inoculum size, the growth rate and the maximum cell density. These values may be evaluated in various ways. The mean growth rate is frequently determined by plotting the natural log of the cell number against time. If a reasonably straight region of the growth curve is selected, the slope gives the mean growth rate, and the y-intercept gives the log of the inoculum size from which the culture apparently developed. This method depends on a subjective estimation of the linear region of the graph. This can be avoided by the use of logistic expressions as outlined below. The mean growth rate can be related to the doubling time by the expression:

Doubling time = ln(2)/mean growth rate

For a very rough idea of the proliferation rate, the doubling time can be determined from the cell yield, the growth time and the inoculum from:

0.3 x time to harvest / $\log_{10}$(yield/inoculum)

This value is always an overestimate of the doubling time, but may be used to give an indication of serious culture problems.

The growth of cultures can be fitted to many mathematical expressions such as the Gompertz equation [159], as well as the Verlhurst equation which will be discussed as an example.

The Verlhurst equation was derived in 1844 to describe the growth of a population under conditions in which a critical factor, such as a nutrient, was limiting. It takes the form:

$$\frac{dN}{dt} = kN(1 - N / N_{\max})$$

which can be integrated to give

$$N_t = \frac{N_0 . e^{kt}}{1 - N_0 / N_{\max}(1 - e^{kt})}$$

where N_t is the number of cells at time t, N_{max} is the maximum number of cells, N_o is the number of cells in the inoculum and k is the mean growth rate of the culture. Experimental values of the cell density as a function of time can be conveniently fitted to this equation (Figure 3) by the method of least squares using the solver function of a spreadsheet such as Excel (Microsoft, Redmond, WA). While the theoretical basis of this equation, limited provision of a factor critical for growth, may not necessarily apply to tissue culture systems as usually grown, other hypotheses, such as generation of a toxic product or a differentiation-inducing factor, give rise to expressions that are indistinguishable under practical conditions.

The Verlhurst expression may be interpreted on the proposition that cells proliferate in an exponential manner, but withdraw from the cell cycle as they reach a maximal density according to the term $(1\text{-}N/N_{max})$. Withdrawal from the cell cycle, however, needs to be interpreted with care because dye dilution experiments demonstrate that fibroblasts proliferate as a single population. It would thus appear that $(1\text{-}N/N_{max})$ rather represents, on average, a greater proportion of the cycle spent in G_1.

Use of the Verlhurst equation assumes a particular relationship between cell density and the drop in proliferation $(1\text{-}N/N_{max})$ that may ultimately be superceded by better theoretical models of the mechanism of the control. However, at this point, the available data are not sufficiently refined to distinguish between such models.

4.3 Viability

Determination of the viability of fibroblast cultures is notoriously difficult. The most rigorous method for determining the number of viable cells in a culture is cloning efficiency, in which cells are plated at a low density such that, as they proliferate, they form discrete clones. Under these conditions, fibroblasts show meager, cell density-dependent plating efficiencies. Thus, although comparative values for viable cell numbers can be obtained, the experiments must be performed with care and the absolute values interpreted with caution. As much as possible, the number of viable clones counted on each plate should lie in the same range (within a factor of about 5) to minimize the effect of density on plating efficiency. This requires plating the cells in a series of dilutions, and then using the appropriate dilution for subsequent data analysis. The colonies should be allowed to grow to at least 16 cells, preferably to more than 64 cells, and then stained (e.g. with Giemsa) and counted. Clearly, in the case of fibroblasts, the technique is only practical with cultures that are pre-senescent by at least 6 population doublings.

Since cloning efficiency is both tedious and provides a poor estimate of viability, other methods have been developed. These depend on membrane permeability or metabolic activity. The simplest utilize dyes to which the cell membrane is impermeable. Thus, trypan blue is excluded by live cells, so that when viewed in a hemacytometer, dead cells are blue and live cells clear and refractile. A practical method is given by Freshney [1]. Fluorescein diacetate [160, 161] and Calcein AM [162] are fluorescein derivatives that are taken up by live cells and hydrolyzed to a product to which the cell membrane is impermeable. These dyes are therefore accumulated by live cells. Calcein AM is generally superior as its loss from the cells is slower than that of fluorescein [163]. DNA binding dyes, such as ethidium bromide [164], propidium iodide, ethidium bromide homodimer I [165-167] and Sytox (Molecular Probes, Eugene, OR) are unable to enter viable cells, but are taken up by dead cells and stain the DNA of the nucleus showing a very large fluorescence enhancement on binding. A number of commercial methods are available based on these dye combinations that provide convenient means to determine culture viability (Molecular Probes). The methods can be applied to FACS and fluorometric techniques [168].

4.4 Extracellular Matrix Synthesis

4.4.1 Collagen deposition

Collagen deposition by fibroblasts has been found to depend on the culture type, ascorbate [14-16, 128], TGF-β_1 [16, 169] and provision of certain amino acids, particularly cystine [170]. Collagen content may be determined by several methods including dye binding, hydroxyproline determination and estimation of cyanogen bromide cleavage products.

4.4.1.1 Sirius red binding

The binding of Sirius Red F3BA (also called Direct Red 80) in saturated picric acid solution has been used widely to estimate fibrotic collagen [171-177]. In these methods, the amount of dye is estimated micro-spectrophotometrically, or eluted with alkali and determined in a spectrophotometer or micro-plate reader. The technique has been adapted using alkali elution to a micro-plate application. In our modification, we solubilize the bound dye with collagenase, which provides more quantitative elution than alkali and is specific for collagen.

The specificity of Sirius Red binding to collagen is based largely on its use as a histological stain. In rat liver with various degrees of cholestatic fibrosis, collagen content measured by Sirius Red binding shows a strong correlation with hydroxyproline content [173, 178]. In addition, histological staining of collagen with Sirius Red is bi-refringent, indicating directional binding related to the orientation of the collagen strands. Sirius Red is known to bind to proteins other than the classical collagens that contain collagen-like triple helices, such as the complement component C1 [178]. Some minor binding to serum albumin has also been found, although control experiments using a bovine serum albumin standard have shown no interference with the assay. We estimate this to represent less than 2% of the collagen signal obtained from typical scaffold-based three-dimensional fibroblast cultures. While the mechanism of dye binding is not clear, comparison of collagen binding by related dyes indicates that the sulfonate groups are unlikely to be important, although they may play a role in bringing the dye into the vicinity of the protein [179]. It has been proposed that the chemical structures providing major binding are the four widely spaced azo groups and planar aromatic rings.

Puchtler et al. point out that Sirius Red F3BA is likely to aggregate in picric acid solution and may not be able to penetrate collagen bundles. It is not dialyzable. Furthermore, they also question the extent to which dye binding may give quantitative estimates of collagen, since the degree of aggregation and binding may depend on the dye concentration. Under our conditions, we use a standard picrosirius red solution that saturates the collagen, and we expect dye binding to be constant. It is likely that the penetration of Sirius Red varies with the packing of collagen bundles. These considerations place a constraint on the application of this technique.

Samples to be tested may be pre-washed with a detergent solution, such as 0.1% Triton X-100, to remove possible interfering cellular proteins, and then incubated with 0.1% Direct Red 80 (Sigma) for 60 minutes with shaking. An adequate volume of dye solution needs to be added. e have found 0.5 mL to be sufficient for samples containing up to about 2.5 mg collagen. At the end of the incubation, the dye solution is removed, the sample washed 3 times with PBS, and a suitable volume (0.5 – 2 mL) of sterile 2 mg/mL bacterial collagenase in PBS (with calcium and magnesium) is added. After incubation for at least 1.5 hours with shaking at 37°C, the concentration of released dye is determined using a micro-plate reader at 540 nm. Standards may be prepared by drying soluble collagen (Sigma) on a woven or knitted scaffold up to about 400μg per cm^2.

4.4.1.2 Hydroxyproline

Hydroxyproline is a unique constituent of collagen required for stabilization of the triple helical structure. Hydroxylation occurs within the cell, much of it during synthesis on the ribosome [180-182], by a reaction that uses ascorbate as a reducing co-factor and decarboxylates 2-oxoglutarate. Hydroxyproline may be determined by reaction with *p*-dimethylaminobenzaldehyde. Its content in mature collagen is 14.3% [183]. Samples containing about 150μg collagen are hydrolyzed overnight at 60°C using papain (1 mg/mL in 0.1M phosphate buffer, 5mM cysteine, 5mM EDTA), followed by overnight treatment at 120°C in 6N HCl. After drying, the residue is taken up in 1mL 0.25M sodium phosphate buffer (pH 6.5) and 50μl samples are oxidized by addition of an equal volume of 67mM Chloramine T in a solution of 1M sodium acetate, 0.15M sodium citrate, 30% isopropanol, pH 6.0 for 20 minutes at room temperature in a 96 well plate. An equal volume of 0.12M *p*-dimethylaminobenzaldehyde in 70% isopropanol containing 1.8M perchloric acid is added and the plate is incubated in a water bath at 60°C for 30 minutes. After allowing the plate to cool for 10 – 15 minutes, the optical density of the solution is determined at 540nm and compared with a range of standards. It should be noted that amino acid analysis of collagen deposited by cultured cells has indicated that the value of 14.3% for the hydroxyproline content of collagen may vary somewhat with source, and this should be verified before using the method routinely [184].

4.4.1.3 Cyanogen bromide peptides

Cyanogen bromide cleaves proteins at methionine residues. Different pepsinized collagens give well-defined peptides that can be separated and quantified by electrophoresis [185], by HPLC [186], or by 2-dimensional electrophoresis [187]. The major advantage of these methods is their ability to distinguish between different types of collagen.

4.4.1.4 Proline incorporation

Since collagen contains a high proportion of proline and hydroxyproline, incorporation of [^{3}H]- or [^{14}C]- proline has long been used to provide an estimate of the relative rates of collagen synthesis under various conditions. It needs to be borne in mind, however, that other proteins contain proline-rich regions (e.g. SH3 interaction sites), so that proline incorporation is by no means unique to collagen synthesis. Careful analysis of collagen secreted by dermal fibroblasts in monolayer culture give values from 38% - 66%, depending on cell density [184]. The specificity of the method may be improved by including digestion with bacterial collagenase as a control. In this case, in contrast to the isolation of cells from tissues (see above), a rather pure grade of collagenase is required, as contaminating proteases may give rise to a falsely high estimation of proline incorporation into collagen.

4.4.1.5 C-terminal propeptide determination

Deposition of collagen involves cleavage of the pro-collagen by a C-terminal pro-collagen peptidase, of which BMP-1 is the prime example [120, 188], with the concomitant release of the C-terminal pro-peptide, which may be measured in medium by ELISA. Commercial kits are available for C-terminal pro-peptides for collagen types I, II and III. Several factors may complicate estimation of deposited collagen from pro-peptide release, including degradation of the pro-peptide and collagen turnover. It is, therefore, important to calibrate the C-terminal pro-peptide release method by comparing deposited collagen with a method such as Sirius red binding or hydroxyproline, thereby ensuring proper stoichiometry. Some experiments using three-dimensional fibroblast cultures have indicated that cumulative pro-peptide release underestimates deposited collagen by a factor that varies with culture conditions, up to about 75%, particularly at high cell densities.

4.4.2 Mathematical description of collagen deposition

Secretion of cellular products has been described by the Leudeking-Piret expression [189], which separates product formation kinetics into growth-associated and non-growth-associated components:

$$dC/dt = \alpha(dN/dt) + \nu N$$

where dC/dt is the rate of collagen production, α is the growth-associated rate constant for collagen formation and ν is the rate constant for non-growth-associated production. However, the best fit of experimental data on the secretion of collagen by fibroblasts to this equation requires a negative value for α, implying either consumption of collagen by growing cells or an inhibitory effect of proliferation. Since neither explanation is satisfactory, an alternative was devised based on the proposal that dividing cells are likely to show reduced collagen secretion. Cell proliferation is well described by the Verlhurst equation as described above, where the term describing the reduction in growth rate with time is interpreted in terms of increased time spent in G_1. It is proposed that during this time, the fibroblasts synthesize and secrete collagen. The proportion of cells in this state is given by N/N_{max}.

Since fibroblasts in the skin, under steady state conditions, deposit no net collagen, it is reasonable to assume that collagen synthesis follows a logistic expression, similar to the Verlhurst equation, ultimately reaching a limiting collagen density. Using these two concepts, an equation for collagen deposition can be derived:

$$dC/dt = DN_t(Nt/N_{\max})[1-(C/C_{\max})]$$

where dC/dt is the rate of collagen deposition, D is a rate constant for collagen deposition, N_t is the cell density at t, C is the collagen density and C_{max} is the maximal collagen density. The term describing the decline in collagen deposition at high extracellular matrix concentrations ($1-C/C_{max}$) is subject to the same reservations as the similar term in the Verlhurst expression, and may be modified with deeper understanding of the mechanism of control of collagen synthesis. This hypothesis provides a satisfactory fit of experimental data, although the very low deposition rate at early times indicates even greater cell density dependence. Introduction of a further cell density dependence term substantially improves the correspondence between the predicted and experimental values at early times. Empirical fitting for collagen deposition by scaffold-based three-dimensional fibroblast cultures in a series of experiments, based on minimizing the coefficient of variation (*i.e.* increasing the weighting of low values) gave a power of 2.6 for cell density. This is greater than the squared function predicted by the derivation described above and implies a further cell-dependent factor. Such a factor might be, for instance, secretion of CTGF, which is highly expressed in three-dimensional cultures, aggregation of pro-collagen [122], or increased local concentrations of pro-collagen and BMP-1. An empirical fit to this equation of data derived from a series of experiments is illustrated in Figure 4. This expression ignores the possible contribution of degradation of newly formed collagen.

In contrast to these results in scaffold-based three-dimensional cultures, monolayer cultures show no dependence of collagen synthesis, expressed on a cellular basis, with cell density [184]. In collagen gel-based three-dimensional cultures, the expression of the collagen gene is repressed [112] and collagen secretion is reduced. This response is more marked in relaxed than in stressed gels.

5. PRACTICAL APPLICATIONS OF DERMAL FIBROBLAST CULTURES

5.1 Studies of the properties of primary cultures

Fibroblasts remain the most convenient source of primary human cells having no known rearrangement in their DNA. For this reason, they are widely used for studies of gene expression in patients with mutations affecting, for instance, DNA repair or aging. In common with many other cell types, fibroblasts can be caused to fuse by treatment with inactivated Sendai virus, or more practically, with polyethylene glycol [190]. This technique has been used to examine the complementation relationships between closely related diseases such as xeroderma pigmentosum [191, 192]. The same approach has been used to demonstrate that the senescent phenotype behaves as though dominant to the proliferative, both in normal fibroblast senescence [193] and in senescence associated with the premature aging disease, Werner syndrome [194]. A detailed description of the technique of polyethylene glycol-induced cell fusion may be found in Freshney [1].

5.2 Fibroblast interactions with other cell types and feeder cultures

While the most obvious differentiated function of fibroblasts is the secretion and maintenance of extracellular matrix, fibroblasts do interact with other cells both through growth factors and potentially through cell contact [195]. Examples of paracrine interactions include the expression by fibroblasts of keratinocyte growth factor [196, 197], the receptor for which is present in the spinous layer of the epidermis but not in the dermis [198], and secretion of IL-8 [199]. Under *in vivo* conditions, these interactions play a role in the control of inflammatory processes and wound repair. It would, therefore, not be surprising if fibroblasts might be used as feeder layers to promote the culture of other cell types.

The first example of the use of fibroblast feeder layers was in cloning HeLa cells [200], and the technique has since been widely exploited for cloning and for the culture of cells such as keratinocytes [201], mammary epithelial cells [202] and tumor cells. In general, the proliferation of the fibroblasts must be prevented so that they do not overgrow the culture of interest. This is usually performed either by lethal radiation with an X-ray or γ-ray source, or by treatment with mitomycin C. Precise conditions have to be determined by experimentation, but incubation overnight with a concentration of 5 μg/mL of mitomycin C has been suggested. As recommended by Freshney, the treated cultures should be sub-cultured and plated at a high density to ensure that no replicative cells remain [1].

An extension of the use of fibroblasts as feeder cultures is the application of three-dimensional fibroblast cultures to support keratinocyte proliferation and differentiation in preparing skin substitutes [101, 102]. Application to other epithelial tissues has been slow but should provide interesting experimental models.

5.3 Tissue engineering of connective tissues other than dermis

With the development of tissue-engineered dermal replacements, attention has turned to the possibility of using similar principles for making other tissues such as ligament and tendon [203, 204]. It is not clear whether the fibroblasts that construct these tissues in vivo are so different from dermal fibroblasts, which are easily obtained, that interconversion is impossible. Several major problems need to be overcome, most notably the development of high mechanical strength, which is not, as yet, characteristic of the extracellular matrix deposited by dermal fibroblasts *in vitro* [205]. More complete discussion of these tissues can be found elsewhere in this volume.

5.4 Therapeutic and testing applications of three-dimensional cultures

Three-dimensional fibroblast cultures have formed the basis of several artificial skin devices, with or without the addition of a keratinocyte-derived epidermal layer, which have been employed in research, testing and therapeutic applications [206]. Collagen gel-based and scaffold-based three-dimensional cultures show quite different properties in terms of cell proliferation, integrin and cytokine expression [119] and need to be chosen carefully. Scaffold-based systems without an epidermal layer have been used for toxicity testing, and with an epidermal layer for toxicity and corrosivity testing and for determining the efficiency of photoprotective agents [207-209]. Similar systems have been developed using the collagen gel-based devices [118, 210, 211].

As tissue engineered equivalents of the dermis, three-dimensional fibroblast cultures have found application as skin grafting materials. Using collagen gel-based cultures, tissue bearing an epidermal layer and designed to closely resemble a skin graft (Apligraf®, Organogenesis, Canton, MA), has been FDA approved for the treatment of venous stasis ulcers and other chronic wounds [212]. Scaffold-based three-dimensional cultures without a keratinocyte layer (TransCyte™, Advanced Tissue Sciences, La Jolla, CA) have been FDA approved for the treatment of full-thickness and partial thickness burns [213-218], and have also been successfully applied to the treatment of diabetic foot ulcers (Dermagraft®, Advanced Tissue Sciences) [216, 219-222]. In this case, it is thought that the improvement in healing results from supplying non-senescent cells, angiogenic growth factors [223, 224] and a substrate that promotes keratinocyte migration [218].

The extracellular matrix laid down by fibroblasts in scaffold-based three-dimensional culture, while consisting largely of collagen, is known to contain fibronectin of the wound healing type, tenascin, decorin, thrombospondin-2, galectin-1 and other proteins [223]. It has been established that the response of fibroblasts to this extracellular matrix differs from that to collagen gels [119], and the possibility exists that this material may prove an interesting substrate for the culture of other cell types.

5.5 Extracellular matrix production

Fibroblasts in scaffold-based three-dimensional cultures produce substantial amounts of extracellular matrix that can be used as a source of human collagen and other molecules. It provides a unique biosynthetic source free from the possibility of viral or prion contamination that may occur in cadaveric sources. This approach to collagen biosynthesis has the advantage over recombinant techniques [225-229] in that it employs cells that already carry the entire, complex machinery involved in collagen synthesis. In the case of type I collagen, this includes genes for both the α_1 and α_2 chains, *hsp*47 [230-233], many enzymes involved in hydroxylation and other modifications [234, 235], secretion, cleavage of the terminal propeptides and deposition [236-238]. While experimental systems that produce recombinant collagen provide much valuable information on the mechanism of synthesis, assembling the entire system is a large and complex undertaking. By contrast, the fibroblasts have the system already optimized.

5.6 Gene therapy applications

Substantial attention has been paid the possibility of using fibroblasts for delivering gene therapy [239-241]. Although expression has been obtained *in vivo* for some weeks [242], therapeutic application of this approach will require careful evaluation. Fibroblasts secrete a variety of growth factors under the control of cellular programs [223], and it is possible that addition of excess growth factor by the application of genetically-modified fibroblasts will greatly affect *in vivo* processes such as wound healing [243]. It is more likely that genetic modification of fibroblasts will find utility in conferring on the cells properties that they would not otherwise have, such as desiccation tolerance [244].

6. CONCLUSION

Fibroblast culture has developed from the basic tissue culture system in which many of the fundamental techniques of the field were established to model systems for the exploration of such fundamental cell biological problems as aging and the special properties of three-dimensional culture. While easily grown, fibroblasts have special properties that need to be borne in mind when using them for routine culture purposes. In particular, human fibroblasts show senescence, and change in properties with time, so that care must be taken to ensure the use of cells at comparable passage number. The cultures remain among the most important model systems for study of the basic properties of cells and provide an important component of tissue engineering systems. Potential applications include development of devices for the treatment of acute and chronic wounds, formation of connective tissues other than skin and delivery of gene products from transfected cells.

ACKNOWLEDGEMENTS

The author would like to thank Susan Edwards, Robert Harding, Mark Baumgartner, Ann Remillard, Frank Zeigler, Ruth Patch, Holly Alexander, Inger Kidd, Andreas Kern and the staff research and development groups at Advanced Tissue Sciences, Inc. for providing experimental data, and Jeffrey Winkelman, Drs. Tony Ratcliffe, David Horwitz and Vera Morhenn for helpful comments on the manuscript.

REFERENCES

1. Freshney RI. (1987) *Culture of Animal Cells. A Manual of Technique.* 2nd Edition ed, Wiley-Liss. New York.
2. Carrel A, (1912) On the permanent life of tissues outside of the organism. *J. Exp. Med.*, 15:516-528.
3. Medawar PB. (1957) *The Uniqueness of the Inidvidual*, Basic Books.
4. Schnieke AE, Kind AJ, Ritchie WA, Mycock K, Scott AR, Ritchie M, Wilmut I, Colman A and Campbell KH, (1997) Human factor IX transgenic sheep produced by transfer of nuclei from transfected fetal fibroblasts. *Science*, 278:2130-2133.

5. Cibelli JB, Stice SL, Golueke PJ, Kane JJ, Jerry J, Blackwell C, Ponce de Leon FA and Robl JM, (1998) Cloned transgenic calves produced from nonquiescent fetal fibroblasts. *Science*, 280:1256-1258.
6. Todaro GJ and Green H, (1963) Quantitative studies of the growth of mouse embryo cells in culture and their development into established cell lines. *J. Cell Biol.*, 17:299-313.
7. Rittling SR, (1996) Clonal nature of spontaneously immortalized 3T3 cells. *Exp Cell Res*, 229:7-13.
8. Ross SM, Joshi R and Frank CB, (1990) Establishment and comparison of fibroblast cell lines from the medial collateral and anterior cruciate ligaments of the rabbit. *In Vitro Cell Dev Biol*, 26:579-584.
9. Falanga V, Zhou LH, Takagi H, Murata H, Ochoa S, Martin TA and Helfman T, (1995) Human dermal fibroblast clones derived from single cells are heterogeneous in the production of mRNAs for α1(1) procollagen and transforming growth factor β1. *J. Invest. Dermatol.*, 105:27-31.
10. Korn JH, Torres D and Downie E, (1984) Clonal heterogeneity in the fibroblast response to mononuclear cell derived mediators. *Arthritis Rheum*, 27:174-179.
11. Goldring SR, Stephenson ML, Downie E, Krane SM and Korn JH, (1990) Heterogeneity in hormone responses and patterns of collagen synthesis in cloned dermal fibroblasts. *J. Clin. Invest.*, 85:798-803.
12. Bayreuther K, Rodemann HP, Francz PI and Maier K, (1988) Differentiation of fibroblast stem cells. *J. Cell Sci.*,:115-130.
13. Korn JH and Downie E, (1989) Clonal interactions in fibroblast proliferation: recognition of self vs. non-self. *J Cell Physiol*, 141:437-440.
14. Grinnell F, Fukamizu H, Pawelek P and Nakagawa S, (1989) Collagen processing, crosslinking, and fibril bundle assembly in matrix produced by fibroblasts in long-term cultures supplemented with ascorbic acid. *Exp Cell Res*, 181:483-491.
15. Ishikawa O, Kondo A, Okada K, Miyachi Y and Furumura M, (1997) Morphological and biochemical analyses on fibroblasts and self-produced collagens in a novel three-dimensional culture. *Br J Dermatol*, 136:6-11.
16. Clark RA, McCoy GA, Folkvord JM and McPherson JM, (1997) TGF-beta 1 stimulates cultured human fibroblasts to proliferate and produce tissue-like fibroplasia: a fibronectin matrix-dependent event. *J Cell Physiol*, 170:69-80.
17. Hehenberger K and Hansson A, (1997) High glucose-induced growth factor resistance in human fibroblasts can be reversed by antioxidants and protein kinase C inhibitors. *Cell Biochem. Funct.*, 15:197-201.
18. Iyer VR, Eisen MB, Ross DT, Schuler G, Moore T, Lee JCF, Trent JM, Staudt LM, Hudson J, Jr., Boguski MS, Lashkari D, Shalon D, Botstein D and Brown PO, (1999) The transcriptional program in the response of human fibroblasts to serum. *Science*, 283:83-87.
19. Chen TR, (1977) In situ detection of mycoplasma contamination in cell cultures by fluorescent Hoechst 33258 stain. *Exp Cell Res*, 104:255-262.
20. Dulbecco R, (1952) Production of plaques in monolayer tissue culture by single particles of an animal virus. *Proc. natl. Acad. Sci. U. S. A.*, 38:747-752.
21. Sanford KK, Earle WR and Likely GD, (1948) The growth in vitro of single isolated tissue cells. *J. natl. Cancer Inst.*, 9:229-246.
22. Wieser RJ, Schutz S, Tschank G, Thomas H, Dienes HP and Oesch F, (1990) Isolation and characterization of a 60-70-kD plasma membrane glycoprotein involved in the contact-dependent inhibition of growth. *J Cell Biol*, 111:2681-2692.

23. Gradl G, Faust D, Oesch F and Wieser RJ, (1995) Density-dependent regulation of cell growth by contactinhibin and the contactinhibin receptor. *Curr Biol*, 5:526-535.
24. Wieser RJ, Baumann CE and Oesch F, (1995) Cell-contact mediated modulation of the sialylation of contactinhibin. *Glycoconj J*, 12:672-679.
25. Trosko JE and Ruch RJ, (1998) Cell-cell communication in carcinogenesis. *Front Biosci*, 3:D208-236.
26. Soler C, Grangeasse C, Baggetto LG and Damour O, (1999) Dermal fibroblast proliferation is improved by beta-catenin overexpression and inhibited by E-cadherin expression. *FEBS Lett*, 442:178-182.
27. Batt DB and Roberts TM, (1998) Cell density modulates protein-tyrosine phosphorylation. *J Biol Chem*, 273:3408-3414.
28. Wieser RJ, Faust D, Dietrich C and Oesch F, (1999) $p16^{INK4}$ mediates contact-inhibition of growth. *Oncogene*, 18:277-281.
29. Dietrich C, Wallenfang K, Oesch F and Wieser R, (1997) Differences in the mechanisms of growth control in contact-inhibited and serum-deprived human fibroblasts. *Oncogene*, 15:2743-2747.
30. Serini G, Bochaton-Piallat ML, Ropraz P, Geinoz A, Borsi L, Zardi L and Gabbiani G, (1998) The fibronectin domain ED-A is crucial for myofibroblastic phenotype induction by transforming growth factor-beta1. *J Cell Biol*, 142:873-881.
31. ffrench CC, Van de Water L, Dvorak HF and Hynes RO, (1989) Reappearance of an embryonic pattern of fibronectin splicing during wound healing in the adult rat. *J Cell Biol*, 109:903-914.
32. Ehrlich HP, Cremona O and Gabbiani G, (1998) The expression of alpha 2 beta 1 integrin and alpha smooth muscle actin in fibroblasts grown on collagen. *Cell Biochem Funct*, 16:129-137.
33. Desmouliere A, Rubbia-Brandt L, Abdiu A, Walz T, Macieira-Coelho A and Gabbiani G, (1992) Alpha-smooth muscle actin is expressed in a subpopulation of cultured and cloned fibroblasts and is modulated by gamma-interferon. *Exp Cell Res*, 201:64-73.
34. Hayflick L and Moorhead PS, (1961) The serial cultivation of human diploid cell strains. *Exp. Cell Res.*, 25:585-621.
35. Goel NS and Ycas M, (1976) The error catastrophe hypothesis and aging. *J Math Biol*, 3:121-147.
36. Holliday R, Huschtscha LI, Tarrant GM and Kirkwood TB, (1977) Testing the commitment theory of cellular aging. *Science*, 198:366-372.
37. Kirkwood TB and Holliday R, (1979) Human cells and the finite lifespan theory. *Adv Exp Med Biol*, 118:35-46.
38. Gupta RS, (1980) Senescence of cultured human diploid fibroblasts. Are mutations responsible? *J Cell Physiol*, 103:209-216.
39. Harley CB and Goldstein S, (1980) Retesting the commitment theory of cellular aging. *Science*, 207:191-193.
40. Angello JC and Prothero JW, (1985) Clonal attenuation in chick embryo fibroblasts. Experimental data, a model and computer simulations. *Cell Tissue Kinet*, 18:27-43.
41. Bayreuther K, Rodemann HP, Hommel R, Dittmann K, Albiez M and Francz PI, (1988) Human skin fibroblasts in vitro differentiate along a terminal cell lineage. *Cell Biology*, 85:5112-5116.
42. Toussaint O, Michiels C, Raes M and Remacle J, (1995) Cellular aging and the importance of energetic factors. *Exp Gerontol*, 30:1-22.
43. Sottile J, Hoyle M and Millis AJ, (1987) Enhanced synthesis of a Mr = 55,000 dalton peptide by senescent human fibroblasts. *J Cell Physiol*, 131:210-217.

44. Seshadri T and Campisi J, (1990) Repression of c-fos transcription and an altered genetic program in senescent human fibroblasts. *Science*, 247:205-209.
45. Porter MB, Pereira-Smith OM and Smith JR, (1990) Novel monoclonal antibodies identify antigenic determinants unique to cellular senescence. *J Cell Physiol*, 142:425-433.
46. Martin M, el Nabout R, Lafuma C, Crechet F and Remy J, (1990) Fibronectin and collagen gene expression during in vitro ageing of pig skin fibroblasts. *Exp Cell Res*, 191:8-13.
47. Liu XT, Stewart CA, King RL, Danner DA, Dell'Orco RT and McClung JK, (1994) Prohibitin expression during cellular senescence of human diploid fibroblasts. *Biochem Biophys Res Commun*, 201:409-414.
48. Tahara H, Hara E, Tsuyama N, Oda K and Ide T, (1994) Preparation of a subtractive cDNA library enriched in cDNAs which expressed at a high level in cultured senescent human fibroblasts. *Biochem Biophys Res Commun*, 199:1108-1112.
49. Palmero I, McConnell B, Parry D, Brookes S, Hara E, Bates S, Jat P and Peters G, (1997) Accumulation of p16INK4a in mouse fibroblasts as a function of replicative senescence and not of retinoblastoma gene status. *Oncogene*, 15:495-503.
50. Miyazaki M, Gohda E, Kaji K and Namba M, (1998) Increased hepatocyte growth factor production by aging human fibroblasts mainly due to autocrine stimulation by interleukin-1. *Biochem Biophys Res Commun*, 246:255-260.
51. Knight JA, (1995) The process and theories of aging. *Ann Clin Lab Sci*, 25:1-12.
52. Pereira-Smith OM and Smith JR, (1988) Genetic analysis of indefinite division in human cells: identification of four complementation groups. *Proc Natl Acad Sci U S A*, 85:6042-6046.
53. Smith JR and Pereira-Smith OM, (1989) Further studies on the genetic and biochemical basis of cellular senescence. *Exp Gerontol*, 24:377-381.
54. Hoehn H, Bryant EM and Martin GM, (1978) The replicative life spans of euploid hybrids derived from short-lived and long-lived human skin fibroblast cultures. *Cytogenet Cell Genet*, 21:282-295.
55. Sugawara O, Oshimura M, Koi M, Annab LA and Barrett JC, (1990) Induction of cellular senescence in immortalized cells by human chromosome 1. *Science*, 247:707-710.
56. Sandhu A, Hubabard K, Kaur GP, Jha K, ozer HL and Athwal RS, (1994) Senescence of immortal human fibroblasts by the introduction of normal human chromosome 6. *Proc. natl. Acad. Sci. U. S. A.*, 91.
57. Hensler PJ, Annab LA, Barrett JC and Pereira-Smith OM, (1994) A gene involved in control of human cellular senescence on human chromosome 1q. *Mol Cell Biol*, 14:2291-2297.
58. Cristofalo VJ, Volker C, Francis MK and Tresini M, (1998) Age-dependent modifications of gene expression in human fibroblasts. *Crit Rev Eukaryot Gene Expr*, 8:43-80.
59. McConnell BB, Starborg M, Brookes S and Peters G, (1998) Inhibitors of cyclin-dependent kinases induce features of replicative senescence in early passage human diploid fibroblasts. *Curr Biol*, 8:351-354.
60. Stein GH, Drullinger LF, Soulard A and Dulic V, (1999) Differential roles for cyclin-dependent kinase inhibitors p21 and p16 in the mechanisms of senescence and differentiation in human fibroblasts. *Mol Cell Biol*, 19:2109-2117.
61. Hanish JP, Yanowitz JL and de Lange T, (1994) Stringent sequence requirements for the formation of human telomeres. *Proceedings of the National Academy of Sciences, U. S. A.*, 91:8861-8865.

62. Kim NW, Piatyszek MA, Prowse KR, Harley CB, West MD, Ho PL, Coviello GM, Wright WE, Weinrich SL and Shay JW, (1994) Specific association of human telomerase activity with immortal cells and cancer. *Science*, 266:2011-2015.
63. Mantell LL and Greider CW, (1994) Telomerase activity in germline and embryonic cells of Xenopus. *EMBO J*, 13:3211-3217.
64. Blasco MA, Funk W, Villeponteau B and Greider CW, (1995) Functional characterization and developmental regulation of mouse telomerase RNA. *Science*, 269:1267-1270.
65. Bickenbach JR, Vormwald-Dogan V, Bachor C, Bleuel K, Schnapp G and Boukamp P, (1998) Telomerase is not an epidermal stem cell marker and is downregulated by calcium. *J Invest Dermatol*, 111:1045-1052.
66. Levy MZ, Allsopp RC, Futcher AB, Greider CW and Harley CB, (1992) Telomere end-replication problem and cell aging. *J Mol Biol*, 225:951-960.
67. Greider CW, (1994) Mammalian telomere dynamics: healing, fragmentation shortening and stabilization. *Curr Opin Genet Dev*, 4:203-211.
68. Allsopp RC, Vaziri H, Patterson C, Goldstein S, Younglai EV, Futcher AB, Greider CW and Harley CB, (1992) Telomere length predicts replicative capacity of human fibroblasts. *Proc Natl Acad Sci U S A*, 89:10114-10118.
69. Filatov L, Golubovskaya V, Hurt JC, Byrd LL, Phillips JM and Kaufmann WK, (1998) Chromosomal instability is correlated with telomere erosion and inactivation of G2 checkpoint function in human fibroblasts expressing human papillomavirus type 16 E6 oncoprotein. *Oncogene*, 16:1825-1838.
70. Bodnar AG, Ouellette M, Frolkis M, Holt S, E., Chiu C-P, Morin GB, Harley CB, Shay JW, Lichteteiner S and Wright WE, (1998) Extension of life-span by introduction of telomerase into normal human cells. *Science*, 279:349-352.
71. Morales CP, Holt SE, Ouellette M, Kaur KJ, Yan Y, Wilson KS, White MA, Wright WE and Shay JW, (1999) Absence of cancer-associated changes in human fibroblasts immortalized with telomerase. *Nat Genet*, 21:115-118.
72. Kiyono T, Foster SA, Koop JI, McDougall JK, Galloway DA and Klingelhutz AJ, (1998) Both Rb/p16^{INK4a} inactivation and telomerase activity are required to immortalize human epithelial cells. *Nature*, 396:84-88.
73. Vogt M, Haggblom C, Yeargin J, Christiansen-Weber T and Haas M, (1998) Independent induction of senescence by p16^{INK4a} and p21^{CIP1} in spontaneously immortalized human fibroblasts. *Cell Growth Differ*, 9:139-146.
74. Medcalf AS, Klein-Szanto AJ and Cristofalo VJ, (1996) Expression of p21 is not required for senescence of human fibroblasts. *Cancer Res*, 56:4582-4585.
75. Wilson VL and Jones PA, (1983) DNA methylation decreases in aging but not in immortal cells. *Science*, 220:1055-1057.
76. Holliday R, (1986) Strong effects of 5-azacytidine on the in vitro lifespan of human diploid fibroblasts. *Exp Cell Res*, 166:543-552.
77. Noble JR, Rogan EM, Neumann AA, Maclean K, Bryan TM and Reddel RR, (1996) Association of extended in vitro proliferative potential with loss of p16^{INK4} expression. *Oncogene*, 13:1259-1268.
78. Perillo NL, Walford RL, Newman MA and Effros RB, (1989) Human T lymphocytes possess a limited in vitro life span. *Exp Gerontol*, 24:177-187.
79. Van Gansen P and Van Lerberghe N, (1988) Potential and limitations of cultivated fibroblasts in the study of senescence in animals. A review on the murine skin fibroblasts system. *Arch Gerontol Geriatr*, 7:31-74.
80. Stevenson AF and Cremer T, (1981) Senescence in vitro and ionising radiations--the human diploid fibroblast model. *Mech Ageing Dev*, 15:51-63.

81. Holliday R, (1991) A re-examination of the effects of ionizing radiation on lifespan and transformation of human diploid fibroblasts. *Mutat Res*, 256:295-302.
82. Tsutsui T, Tanaka Y, Matsudo Y, Hasegawa K, Fujino T, Kodama S and Barrett JC, (1997) Extended lifespan and immortalization of human fibroblasts induced by X- ray irradiation. *Mol Carcinog*, 18:7-18.
83. Li S, Mawal-Dewan M, Cristofalo VJ and Sell C, (1998) Enhanced proliferation of human fibroblasts, in the presence of dexamethasone, is accompanied by changes in $p21^{Waf1/Cip1/Sdi1}$ and the insulin-like growth factor type 1 receptor. *J Cell Physiol*, 177:396-401.
84. Fairweather DS, Fox M and Margison GP, (1987) The in vitro lifespan of MRC-5 cells is shortened by 5-azacytidine-induced demethylation. *Exp Cell Res*, 168:153-159.
85. Shelton DN, Chang E, Whittier PS, Choi D and Funk WD, (1999) Microarray analysis of replicative senescence. *Curr Biol*, 9:939-945.
86. Ly DH, Lockhart DJ, Lerner RA and Schultz PG, (2000) Mitotic misregulation and human aging. *Science*, 287:2486-2492.
87. West MD, Pereira SO and Smith JR, (1989) Replicative senescence of human skin fibroblasts correlates with a loss of regulation and overexpression of collagenase activity. *Exp Cell Res*, 184:138-147.
88. Millis AJ, Hoyle M, McCue HM and Martini H, (1992) Differential expression of metalloproteinase and tissue inhibitor of metalloproteinase genes in aged human fibroblasts. *Exp Cell Res*, 201:373-379.
89. Black AF, Berthod F, L'Heureux N, Germain L and Auger FA, (1998) In vitro reconstruction of a human capillary-like network in a tissue- engineered skin equivalent. *Faseb J*, 12:1331-1340.
90. Burke EM, Horton WE, Pearson JD, Crow MT and Martin GR, (1994) Altered transcriptional regulation of human interstitial collagenase in cultured skin fibroblasts from older donors. *Exp Gerontol*, 29:37-53.
91. Applegate DR, Liu K and Mansbridge J, (1999) Practical considerations for large-scale cryopreservation of a tissue engineered human dermal replacement. *Advances in Heat and Mass Transfer in Biotechnology*.
92. McCormick JJ and Maher VM, (1988) Towards an understanding of the malignant transformation of diploid human fibroblasts. *Mutat Res*, 199:273-291.
93. Mukherji B, MacAlister TJ, Guha A, Gillies CG, Jeffers DC and Slocum SK, (1984) Spontaneous in vitro transformation of human fibroblasts. *J. nat. Cancer. Inst.*, 73:583-593.
94. Honda T, Sadamori N, Oshimura M, Horikawa I, Omura H, Komatsu K and Watanabe M, (1996) Spontaneous immortalization of cultured skin fibroblasts obtained from a high-dose atomic bomb survivor. *Mutat Res*, 354:15-26.
95. Milo GE and Casto BC, (1986) Conditions for transformation of human fibroblast cells: an overview. *Cancer Lett*, 31:1-13.
96. Mayne LV, Priestley A, James MR and Burke JF, (1986) Efficient immortalization and morphological transformation of human fibroblasts by transfection with SV40 DNA linked to a dominant marker. *Exp Cell Res*, 162:530-538.
97. Vertino PM, Issa JP, Pereira-Smith OM and Baylin SB, (1994) Stabilization of DNA methyltransferase levels and CpG island hypermethylation precede SV40-induced immortalization of human fibroblasts. *Cell Growth Differ*, 5:1395-1402.
98. Ray FA and Kraemer PM, (1992) Frequent deletions in nine newly immortal human cell lines. *Cancer Genet Cytogenet*, 59:39-44.

99. Small MB, Hubbard K, Pardinas JR, Marcus AM, Dhanaraj SN and Sethi KA, (1996) Maintenance of telomeres in SV40-transformed pre-immortal and immortal human fibroblasts. *J Cell Physiol*, 168:727-736.
100. Bell E, Ivarsson G and Merrill C, (1979) Production of a tissue-like structure by contraction of collagen lattices by human fibroblasts of different proliferative potential in vitro. *Proc. Natl. Acad. Sci. USA*, 76:1274-1278.
101. Bell E, Ehrlich HP, Buttle DJ and Nakatsuji T, (1981) Living tissue formed in vitro and accepted as skin-equivalent tissue of full thickness. *Science*, 211:1052-1054.
102. Bell E, Sher S, Hull B, Merrill C, Rosen S, Chamson A, Asselineau D, Dubertret L, Coulomb B, Lapiere C, Nusgens B and Neveux Y, (1983) The reconstitution of living skin. *J Invest Dermatol*, 81:2s-10s.
103. Landeen LK, Zeigler FC, Halberstadt C, Cohen R and Slivka SR, (1992) Characterization of a human dermal replacement. *Wounds*, 4:167-175.
104. Slivka SR, Landeen LK, Zeigler F, Zimber MP and Bartel RL, (1993) Characterization, barrier function, and drug metabolism of an in vitro skin model. *J Invest Dermatol*, 100:40-46.
105. Bell E, Rosenberg M, Kemp P, Gay R, Green GD, Muthukumaran N and Nolte C, (1991) Recipes for reconstituting skin. *J Biomech Eng*, 113:113-119.
106. He Y and Grinnell F, (1994) Stress relaxation of fibroblasts activates a cyclic AMP signaling pathway. *J Cell Biol*, 126:457-464.
107. Nakagawa S, Pawelek P and Grinnell F, (1989) Extracellular matrix organization modulates fibroblast growth and growth factor responsiveness. *Exp Cell Res*, 182:572-582.
108. Lin YC and Grinnell F, (1993) Decreased level of PDGF-stimulated receptor autophosphorylation by fibroblasts in mechanically relaxed collagen matrices. *J Cell Biol*, 122:663-672.
109. Nusgens B, Merrill C, Lapiere C and Bell E, (1984) Collagen biosynthesis by cells in a tissue equivalent matrix in vitro. *Collagen Rel. Res.*, 4:351-364.
110. Nakagawa S, Pawelek P and Grinnell F, (1989) Long-term culture of fibroblasts in contracted collagen gels: effects on cell growth and biosynthetic activity. *J Invest Dermatol*, 93:792-798.
111. Mochitate K, Pawelek P and Grinnell F, (1991) Stress relaxation of contracted collagen gels: disruption of actin filament bundles, release of cell surface fibronectin, and down-regulation of DNA and protein synthesis. *Exp Cell Res*, 193:198-207.
112. Langholz O, Röckel D, Mauch C, Kozlowska E, Bank I, Krieg T and Eckes B, (1995) Collagen and collagenase gene expression in three-dimensional collagen lattices are differentially regulated by a1b1 and a2b1 integrins. *J. Cell Biol.*, 131:1903-1915.
113. Eckes B, Krieg T, Nusgens BV and Lapière CM, (1995) In vitro reconstituted skin as a tool for biology, pharmacology and therapy: a review. *Wound Rep. Regen.*, 3:248-257.
114. Xu J, Zutter MM, S.A. S and Clark RAF, (1995) A three-dimensional collagen lattice activates NF-κB in human fibroblasts: role in integrin α_2 gene expression and tissue remodeling. *The Journal of Cell Biology*, 140:709-719.
115. Sher AE, Hull BE, Rosen S, Church D, Friedman L and Bell E, (1983) Acceptance of allogeneic fibroblasts in skin equivalent transplants. *Transplantation*, 36:552-557.
116. Bell E, Sher S and Hull B, (1984) The living skin-equivalent as a structural and immunological model in skin grafting. *Scanning Electron Microscopy*, IV:1957-1962.
117. Grinnell F, Takashima A and Lamke-Seymour C, (1986) Morphological appearance of epidermal cells cultured on fibroblast-reorganized collagen gels *Cell Tissue Res*, 246:13-21.

118. Parenteau NL, Nolte CM, Bilbo P, Rosenberg M, Wilkins LM, Johnson EW, Watson S, Mason VS and Bell E, (1991) Epidermis generated in vitro: practical considerations and applications. *J Cell Biochem*, 45:245-251.
119. Mansbridge JN, Pinney ER, Liu K and Kern A, (1999) Comparison of fibroblast properties in scaffold-based and collagen gel three-dimensional culture systems. *J. Invest. Dermatol.*, 112:536.
120. Li SW, Sieron AL, Fertala A, Hojima Y, Arnold WV and Prockop DJ, (1996) The C-proteinase that processes procollagens to fibrillar collagens is identical to the protein previously identified as bone morphogenic protein-1. *Proc Natl Acad Sci U S A*, 93:5127-5130.
121. Kessler E, Takahara K, Biniaminov L, Brusel M and Greenspan DS, (1996) Bone morphogenetic protein-1: the type I procollagen C-proteinase. *Science*, 271:360-362.
122. Hojima Y, Behta B, Romanic AM and Prockop DJ, (1994) Cleavage of type I procollagen by C- and N-proteinases is more rapid if the substrate is aggregated with dextran sulfate or polyethylene glycol. *Anal Biochem*, 223:173-180.
123. Fruton JS and Simmonds S. (1959) *General Biochemistry*. 2nd edition ed, John Wiley. New York.
124. Kivirikko KI, Myllyla R and Pihlajaniemi T, (1989) Protein hydroxylation: prolyl 4-hydroxylase, an enzyme with four cosubstrates and a multifunctional subunit. *FASEB J*, 3:1609-1617.
125. Pihlajaniemi T, Myllyla R and Kivirikko KI, (1991) Prolyl 4-hydroxylase and its role in collagen synthesis. *J Hepatol*, 13 Suppl 3:S2-7.
126. Murad S, Grove D, Lindberg KA, Reynolds G, Sivarajah A and Pinnell SR, (1981) Regulation of collagen synthesis by ascorbic acid. *Proc Natl Acad Sci U S A*, 78:2879-2882.
127. Russell SB, Russell JD and Trupin KM, (1981) Collagen synthesis in human fibroblasts: effects of ascorbic acid and regulation by hydrocortisone. *J Cell Physiol*, 109:121-131.
128. Schwarz RI, Mandell RB and Bissell MJ, (1981) Ascorbate induction of collagen synthesis as a means for elucidating a mechanism of quantitative control of tissue-specific function. *Mol Cell Biol*, 1:843-853.
129. Lyons BL and Schwarz RI, (1984) Ascorbate stimulation of PAT cells causes an increase in transcription rates and a decrease in degradation rates of procollagen mRNA. *Nucleic Acids Res*, 12:2569-2579.
130. Pinnel SR, Murad S and Darr D, (1987) Induction of collagen synthesis by ascorbic acid. A possible mechanism. *Arch Dermatol*, 123:1684-1686.
131. Bradham DM, Igarashi A, Potter RL and Grotendorst GR, (1991) Connective tissue growth factor: a cysteine-rich mitogen secreted by human vascular endothelial cells is related to the SRC-induced immediate early gene product CEF-10. *J Cell Biol*, 114:1285-1294.
132. Grotendorst GR, Okochi H and Hayashi N, (1996) A novel transforming growth factor beta response element controls the expression of the connective tissue growth factor gene. *Cell Growth Differ*, 7:469-480.
133. Lauffenburger DA, Oehrtman GT, Walker L and Wiley HS, (1998) Real-time quantitative measurement of autocrine ligand binding indicates that autocrine loops are spatially localized. *Proc Natl Acad Sci U S A*, 95:15368-15373.
134. Grotendorst GR, (1997) Connective tissue growth factor: a mediator of TGF-beta action on fibroblasts. *Cytokine Growth Factor Rev*, 8:171-179.
135. Angello JC, Pendergrass WR, Norwood TH and Prothero J, (1989) Cell enlargement: one possible mechanism underlying cellular senescence. *J Cell Physiol*, 140:288-294.

136. Boivin A, Vendrely R and Vendrely C, (1948) L'acide désoxyribonucléique du noyau cellulaire, dépositaire des caractères héréditaires; argument d'ordre analytique. *Comptes rendues à L'Academie Français*, 226:1061-1063.
137. Mirsky AE and Ris H, (1949) Variable and constant components of chromosomes. *Nature*, 163:666-667.
138. McHale AP and McHale L, (1988) Use of tetrazolium based colorimetric assay in assessing phtoradiation therapy in vitro. *Cancer Lett.*, 41:315-321.
139. van de Loosdrecht AA, Nennie E, Ossenkoppele GJ, Beelen RH and Langenhuijsen MM, (1991) Cell mediated cytotoxicity against U 937 cells by human monocytes and macrophages in a modified colorimetric MTT assay. A methodological study. *J Immunol Methods*, 141:15-22.
140. Smith MD, Barbenel JC, Courtney JM and Grant MH, (1992) Novel quantitative methods for the determination of biomaterial cytotoxicity. *Int J Artif Organs*, 15:191-194.
141. van de Loosdrecht AA, Ossenkoppele GJ, Beelen RH, Broekhoven MG and Langenhuijsen MM, (1992) Role of interferon gamma and tumour necrosis factor alpha in monocyte-mediated cytostasis and cytotoxicity against a human histiocytic lymphoma cell line. *Cancer Immunol Immunother*, 34:393-398.
142. Bahbouth E, Siwek B, De P, Gillet MC, Sabbioni E and Bassleer R, (1993) Effects of trace metals on mouse B16 melanoma cells in culture. *Biol Trace Elem Res*, 36:191-201.
143. Lomonte B, Gutierrez JM, Romero M, Nunez J, Tarkowski A and Hanson LA, (1993) An MTT-based method for the in vivo quantification of myotoxic activity of snake venoms and its neutralization by antibodies. *J Immunol Methods*, 161:231-237.
144. Behl C, Davis JB, Lesley R and Schubert D, (1994) Hydrogen peroxide mediates amyloid b protein toxicity. *Cell*, 77:817-827.
145. Musser DA and Oseroff AR, (1994) The use of tetrazolium salts to determine sites of damage to the mitochondrial electron transport chain in intact cells following in vitro photodynamic therapy with photofrin II. *Photochem. Photobiol.*, 59:621-626.
146. Mosman T, (1983) Rapid colorimetric assay for cellular growth and survival: application to proliferation and cytotoxicity assays. *J. immunol. Methods*, 65:55-63.
147. Monner DA, (1988) An assay for growth of mouse bone marrow cells in microtiter liquid culture using the tetrazolium salt MTT, and its application to studies of myelopoiesis. *Immunol Lett*, 19:261-268.
148. Hahm HA and Ip MM, (1990) Primary culture of normal rat mammary epithelial cells within a basement membrane matrix. I. Regulation of proliferation by hormones and growth factors. *In Vitro Cell Dev Biol*, 26:791-802.
149. Schadendorf D, Worm M and Czarnetzki BM, (1993) Determination of granulocyte/macrophage-colony-stimulating factor secretion by human melanoma cells and its effects on human melanoma cell proliferation. *J Cancer Res Clin Oncol*, 119:501-503.
150. Slater TF, (1963) Studies on a succinate-neotetrazolium reductase system of rat liver II. Points of coupling with the respiratory chain. *Biochim. biophs. Acta*, 77:365-382.
151. Berridge MV and Tan AS, (1993) Characterization of the cellular reduction of 3-(4,5-dimethylthiazol)-2,5-diphenyltetrazolium bromide (MTT): subcellular localization, substrate dependence, and involvement of mitochondrial electron transport in MTT reduction. *Arch. Biochem. Biophys.*, 303:474-482.
152. Berridge MV, Tan AS and Hilton CJ, (1993) Cyclic adenosine monophosphate promotes cell survival and retards apoptosis in a factor-dependent bone marrow-derived cell line. *Exp Hematol*, 21:269-276.

153. Vistica DT, Skehan P, Scudiero D, Monks A, Pittman A and Boyd MR, (1991) Tetrazolium-based assays for cellular viability: a critical examination of selected parameters affecting formazan production. *Cancer Res.*, 51:2515-1520.
154. Slezak SE and Horan PK, (1989) Fluorescent in vivo tracking of hematopoietic cells. Part I. Techical considerations. *Blood*, 74:2172-2177.
155. Samlowski WE, Robertson BA, Draper BK, Prystas E and McGregor JR, (1991) Effects of supravital fluorochromes used to analyze the in vivo homing of murine lymphocytes on cellular function. *J. Immunol. Meth.*, 144:101-115.
156. Hugo P, Kappler JW, Godfrey DI and Marrack PC, (1992) A cell line that can induce thymocyte positive selection. *Nature*, 360:679-682.
157. De Clerck LS, Bridts CH, Mertens AM, Moens MM and Stevens WJ, (1994) Use of fluorescent dyes in the determination of adherence of human leucocytes to endothelial cells and the effect of fluorochromes on cellular function. *J. Immunol. Meth.*, 172:115-124.
158. Horan PK and Slezak SE, (1989) Stable cell membrane labelling. *Nature*, 340:167-168.
159. Demicheli R, Foroni R, Ingrosso A, Pratesi G, Soranzo and Tortoreto M, (1989) An exponential-gompertzian description of lo vo cell tumor growth from in vivo and in vitro data. *Cancer Research*, 49:6543-6546.
160. Rotman B and Papermaster BW, (1966) Membrane properties of living mammalian cells as studied by enzymatic hydrolysis of fluorogenic esters. *Proc. natl. Acad Sci. U.S.A.*, 55:134-141.
161. Martel JL, Jaramillo S, Allen JFH and Rubinstein P, (1974) Serology for automated cytotoxicity assays. *Vox. Sang.*, 27:13-20.
162. Papadopoulos NG, Dedoussis GVZ, Spanakos G, Gritzapis AD, Baxevanis CN and Papamichail M, (1994) An improved fluorescence assay for the determination of lymphocyte-mediated cytotoxicity using flow cytometry. *J. Immunol. Meth.*, 177:101-111.
163. Lichtenfels R, Biddison WE, Schulz H, Vogt AB and Martin R, (1994) CARE LASS (calcein-release-assay), an improved fluorescence-based test system to measure cytotoxic T lymphocyte activity. *J. Immunol. Meth.*, 172:227-239.
164. Beletsky IP and Umansky SR, (1990) A new assay for cell death. *J. Immunol. Meth.*, 134:201-205.
165. Glazer AN, Peck K and Mathies RA, (1990) A stable double-stranded DNA-ethidium homodimer complex: application to picogram fluorescence detection of DNA in agarose gels. *Proc. natl. Acad. Sci., U.S.A.*, 87:3851-3855.
166. Lévesque A, Paquet A and Pagé M, (1995) Improved fluorescent bioassay for hte detection of tumor necrosis factor. *J. Immunol. Meth.*, 178:71-76.
167. Rye HS and Galzer AN, (1995) Interaction of of dimeric intercalating dyes with single-stranded DNA. *Nucl. Acids Res.*, 23:1215-1222.
168. Mansbridge J, Liu K, Patch R, Symons K and Pinney E, (1998) Three-dimensional fibroblast culture implant for the treatment of diabetic foot ulcers: metabolic activity and therapeutic range. *Tissue Eng*, 4:403-414.
169. Roberts AB, Sporn MB, Assoian RK, Smith JM, Roche NS, Wakefield LM, Heine UI, Liotta LA, Falanga V, Kehrl JH and Fauci AS, (1986) Transforming growth factor type beta: rapid induction of fibrosis and angiogenesis *in vivo* and stimulation of collagen formation *in vitro*. *Proc Natl Acad Sci U S A*, 83:4167-4171.
170. Rishikof DC, Kuang PP, Poliks C and Goldstein RH, (1998) Regulation of type I collagen mRNA in lung fibroblasts by cystine availability. *Biochem J*, 331:417-422.
171. Bedossa P, Lemaigre G and Bacci JM, E., (1989) Quantitative estimation of the collagen content in normal and pathologic pancreas tissue. *Digestion*, 44:7-13.

172. Finkelstein I, Trope GE, Basu PK, Hasany SM and Hunter WS, (1990) Quantitative analysis of collagen content and amino acids in trabecular meshwork. *Br J Ophthalmol*, 74:280-282.
173. James J, Bosch KS, Aronson DC and Houtkooper JM, (1990) Sirius red histophotometry and spectrophotometry of sections in the assessment of the collagen content of liver tissue and its application in growing rat liver. *Liver*, 10:1-5.
174. Valderrama R, Navarro S, Campo E, Camps J, Gimenez A, Pares A and Caballeria J, (1991) Quantitative measurement of fibrosis in pancreatic tissue. Evaluation of a colorimetric method. *Int J Pancreatol*, 10:23-29.
175. Campbell SE, Diaz AA and Weber KT, (1992) Fibrosis of the human heart and systemic organs in adrenal adenoma. *Blood Press*, 1:149-156.
176. Campbell SE, Janicki JS, Matsubara BB and Weber KT, (1993) Myocardial fibrosis in the rat with mineralocorticoid excess. Prevention of scarring by amiloride. *Am J Hypertens*.
177. Nohlgard C, Rubio CA, Kock Y and Hammar H, (1993) Liver fibrosis quantified by image analysis in methotrexate-treated patients with psoriasis. *J Am Acad Dermatol*, 28:40-45.
178. Walsh BJ, Thornton SC, Penny R and Breit SN, (1992) Microplate reader-based quantitation of collagens. *Analyt. Biochem.*, 203:187-190.
179. Puchtler H, Meloan SN and Waldrop FS, (1988) Are picro-dye reactions for collagens quantitative? Chemical and histochemical considerations. *Histochemistry*, 88:243-256.
180. Bates CJ, Prynne CJ and Levene CI, (1972) Ascorbate-dependent differences in the hydroxylation of proline and lysine in collagen synthesized by 3T6 fibroblasts in culture. *Biochim Biophys Acta*, 278:610-616.
181. Veis A and Brownell AG, (1977) Triple-helix formation on ribosome-bound nascent chains of procollagen: deuterium-hydrogen exchange studies. *Proc Natl Acad Sci U S A*, 74:902-905.
182. Eleftheriades EG, Ferguson AG, Spragia ML and Samarel AM, (1995) Prolyl hydroxylation regulates intracellular procollagen degradation in cultured rat cardiac fibroblasts. *J Mol Cell Cardiol*, 27:1459-1473.
183. Woessner JF, (1961) The determination of hydroxyproline in tissue and protein samples containing small amounts of this amino acid. *Archives of Biochem, Biophysics*, 93: 440-447.
184. Freiberger H, Grove D, Sivarajah A and Pinnell SR, (1980) Procollagen I synthesis in human skin fibroblasts: effect on culture conditions on biosynthesis. *J Invest Dermatol*, 75:425-430.
185. O'Driscoll SW, Commisso CN and Fitzsimmons JS, (1995) Type II collagen quantification in experimental chondrogenesis. *Osteoarthritis and Cartilage*, 3:197-203.
186. Harwood F and Amiel D, (1987) Semiquantitative HPLC analysis of types I and III collagen in soft tissues. *Liquid Chromatography*, 4:122-124.
187. Lovvorn HN, 3rd, Cheung DT, Nimni ME, Perelman N, Estes JM and Adzick NS, (1999) Relative distribution and crosslinking of collagen distinguish fetal from adult sheep wound repair. *J Pediatr Surg*, 34:218-223.
188. Prockop DJ, Sieron AL and Li SW, (1998) Procollagen N-proteinase and procollagen C-proteinase. Two unusual metalloproteinases that are essential for procollagen processing probably have important roles in development and cell signaling. *Matrix Biol*, 16:399-408.
189. Bailey JE and Ollis DE. (1986)*Biochemical Engineering Fundamentals*. 2nd ed, McGraw Hill. New York.

190. Pontecorvo G, (1975) Production of mammalian somatic cell hybrids by means of polyethylene glycol treatment. *Somatic Cell Genet*, 1:397-400.
191. Kraemer KH, Coon HG, Petinga RA, Barrett SF, Rahe AE and Robbins JH, (1975) Genetic heterogeneity in xeroderma pigmentosum: complementation groups and their relationship to DNA repair rates. *Proc Natl Acad Sci U S A*, 72:59-63.
192. Bootsma D, De Weerd-Kastelein EA, Kleijer WJ and Keyzez W, (1975) Genetic complementation analysis of xeroderma pigmentosum. *Basic Life Sci*, 5B:725-728.
193. Stein GH, Namba M and Corsaro CM, (1985) Relationship of finite proliferative lifespan, senescence, and quiescence in human cells. *J Cell Physiol*, 122:343-349.
194. Salk D, Au K, Hoehn H, Stenchever MR and Martin GM, (1981) Evidence of clonal attenuation, clonal succession, and clonal expansion in mass cultures of aging Werner's syndrome skin fibroblasts. *Cytogenet Cell Genet*, 30:108-117.
195. Smola H, Thiekotter G and Fusenig NE, (1993) Mutual induction of growth factor gene expression by epidermal-dermal cell interaction. *J Cell Biol*, 122:417-429.
196. Werner S, Peters KG, Longaker MT, Fuller-Pace F, Banda MJ and Williams LT, (1992) Large induction of keratinocyte growth factor in the dermis during wound healing. *Proc. natl. acad. Sci. U.S.A.*, 89:6896-6900.
197. Brauchle M, Angermeyer K, Hubner G and Werner S, (1994) Large induction of keratinocyte growth factor expression by serum growth factors and pro-inflammatory cytokines in cultured fibroblasts. *Oncogene*, 9:3199-3204.
198. LaRochelle WJ, Dirsch OR, Finch PW, Cheon HG, May M, Marchese C, Pierce JH and Aaronson SA, (1995) Specific receptor detection by a functional keratinocyte growth factor-immunoglobulin chimera. *J Cell Biol*, 129:357-366.
199. Schröder JM, Sticherling M, Henneicke HH, Preissener WC and Christophers E, (1990) IL-la or Tumor Necrosis Factor-a Stimulate Release of Three NAP-1/IL-8-Related Neutrophil Chemotactic Proteins in Human Dermal Fibroblasts. *The Journal of Immunology*, 144:2223-2232.
200. Puck TT and Marcus PI, (1955) A rapid method for viable cell titration and clone production in tissue culture: the use of X-irradiated cells to supply conditioning factors. *Proc. natl. Acad. Sci U. S. A.*, 41:432-437.
201. Rheinwald JG and Green H, (1975) Serial cultivation of strains of human epidermal keratinocytes: the formation of keratinizing colonies from single cells. *Cell*, 6:331-343.
202. Ehmann UK, Peterson WD, Jr. and Misfeldt DS, (1984) To grow mouse mammary epithelial cells in culture. *J Cell Biol*, 98:1026-1032.
203. Lin VS, Lee MC, O'Neal S, McKean J and Sung KL, (1999) Ligament tissue engineering using synthetic biodegradable fiber scaffolds. *Tissue Eng*, 5:443-452.
204. Carpenter JE, Thomopoulos S and Soslowsky LJ, (1999) Animal models of tendon and ligament injuries for tissue engineering applications. *Clin Orthop*,:S296-311.
205. Frank C, Shrive N, Hiraoka H, Nakamura N, Kaneda Y and Hart D, (1999) Optimisation of the biology of soft tissue repair. *J Sci Med Sport*, 2:190-210.
206. Parenteau N and Naughton G, (1999) Skin: the first tissue-engineered products. *Scientific American*, April:83-85.
207. Naughton GK, Jacob L and Naughton BA, (1989) A physiological skin model for *in vitro* toxicity studies., in *In Vitro Toxicology: New Directions*, AM Goldberg, Editor. Mary Ann Liebert, Inc.: New York. p. 183-189.
208. Triglia D, Sherard Braa S, Yonan C and Naughton GK, (1991) In vitro toxicity of various classes of test agents using the neutral red assay on a human three-dimensional physiological skin model. *In Vitro Cellular and Developmental Biology*, 27A:239-244.
209. Whalen E, Donnelly TA, Naughton G and Rheins LA, (1994) The development of three-dimensional in vitro human tissue models. *Hum Exp Toxicol*, 13:853-859.

210. Johnson EW, Meunier SF, Roy CJ and Parenteau NL, (1992) Serial cultivation of normal human keratinocytes: a defined system for studying the regulation of growth and differentiation. *In Vitro Cell Dev Biol*, 28A:429-435.
211. Nolte CJ, Oleson MA, Bilbo PR and Parenteau NL, (1993) Development of a stratum corneum and barrier function in an organotypic skin culture. *Arch Dermatol Res*, 285:466-474.
212. Sabolinski ML, Alvarez O, Auletta M, Mulder G and Parenteau NL, (1996) Cultured skin as a 'smart material' for healing wounds: experience in venous ulcers. *Biomaterials*, 17:311-320.
213. Hansbrough JF, Morgan J, Greenleaf G and Underwood J, (1994) Development of a temporary living skin replacement composed of human neonatal fibroblasts cultured in Biobrane, a synthetic dressing material. *Surgery*, 115:633-644.
214. Hansbrough JF, (1996) Cultured epidermal and dermal skin replacements, in *Yearbook of Cell and Tissue Transplantation*, RP Lanza and WL Chick, Editors. Kluwer Academic Publishers: Netherlands. p. 205-219.
215. Hansbrough J, D. Dore', J. Noordenbos, W. Hansbrough, (1997) Dermagraft-TC as a Biologic Covering for Partial-Thickness Burns. *Burn Care & Rehabilitation*,:S104.
216. Purdue GF, Hunt JL, Still JM, Jr., Law EJ, Herndon DN, Goldfarb IW, Schiller WR, Hansbrough JF, Hickerson WL, Himel HN, Kealey GP, Twomey J, Missavage AE, Solem LD, Davis M, Totoritis M and Gentzkow GD, (1997) A multicenter clinical trial of a biosynthetic skin replacement, Dermagraft-TC, compared with cryopreserved human cadaver skin for temporary coverage of excised burn wounds. *J Burn Care Rehabil*, 18:52-57.
217. Hansbrough JF, Mozingo DW, Kealey GP, Davis M, Gidner A and Gentzkow GD, (1997) Clinical trials of a biosynthetic temporary skin replacement, Dermagraft-Transitional Covering, compared with cryopreserved human cadaver skin for temporary coverage of excised burn wounds. *J Burn Care Rehabil*, 18:43-51.
218. Krejci-Papa NC, Hoang A and Hansbrough JF, (1999) Fibroblast sheets enable epithelialization of wounds that do not support keratinocyte migration. *Tissue Eng*, 5:555-562.
219. Gentzkow GD, Iwasaki SD, Hershon KS, Mengel M, Prendergast JJ, Ricotta JJ, Steed DP and Lipkin S, (1996) Use of dermagraft, a cultured human dermis, to treat diabetic foot ulcers. *Diabetes Care*, 19:350-354.
220. Naughton GK, Mansbridge JN and Gentzkow G, (1997) A metabolically active human dermal replacement for the treatment of diabetic foot ulcers. *Artificial Organs*, 21:1203-1210.
221. Pollak RA, Edington H, Jensen JL, Kroeker RO and Gentzkow GD, (1997) A human dermal replacement for the treatment of diabetic foot ulcers. *Wounds*, 9:175-183.
222. Gentzkow GD, Jensen LJ, Pollak RA, Kroeker RO, Lerner JM, Lerner M, Iwasaki SD and Group TDDUS, (1999) Improved healing of diabetic foot ulcers after grafting with a living human dermal replacement. *Wounds*, 11:77-84.
223. Mansbridge JN, Liu K, Pinney RE, Patch R, Ratcliffe A and Naughton GK, (1999) Growth Factors Secreted by Fibroblasts. *Diabetes, Obesity and Metabolism*, 1:265-279.
224. Pinney E, Liu K, Sheeman B and Mansbridge J, (2000) Human three-dimensional fibroblast cultures express angiogenic activity. *J Cell Physiol*, 183:74-82.
225. Vuorela A, Myllyharju J, Nissi R, Pihlajaniemi T and Kivirikko KI, (1997) Assembly of human prolyl 4-hydroxylase and type III collagen in the yeast *Pichia pastoris*: formation of a stable enzyme tetramer requires coexpression with collagen and assembly of a stable collagen requires coexpression with prolyl 4-hydroxylase. *Embo J*, 16:6702-6712.

226. Keizer-Gunnink I, Vuorela A, Myllyharju J, Pihlajaniemi T, Kivirikko KI and Veenhuis M, (2000) Accumulation of properly folded human type III procollagen molecules in specific intracellular membranous compartments in the yeast *Pichia pastoris*. *Matrix Biol*, 19:29-36.
227. Vuorela A, Myllyharju J, Pihlajaniemi T and Kivirikko KI, (1999) Coexpression with collagen markedly increases the half-life of the recombinant human prolyl 4-hydroxylase tetramer in the yeast *Pichia pastoris*. *Matrix Biol*, 18:519-522.
228. Alvares K, Siddiqui F, Malone J and Veis A, (1999) Assembly of the type 1 procollagen molecule: selectivity of the interactions between the alpha 1(I)- and alpha 2(I)-carboxyl propeptides. *Biochemistry*, 38:5401-5411.
229. Vaughn PR, Galanis M, Richards KM, Tebb TA, Ramshaw JA and Werkmeister JA, (1998) Production of recombinant hydroxylated human type III collagen fragment in *Saccharomyces cerevisiae*. *DNA Cell Biol*, 17:511-518.
230. Sauk JJ, Smith T, Norris K and Ferreira L, (1994) Hsp47 and the translation-translocation machinery cooperate in the production of alpha 1(I) chains of type I procollagen. *J Biol Chem*, 269:3941-3946.
231. Hu G, Gura T, Sabsay B, Sauk J, Dixit SN and Veis A, (1995) Endoplasmic reticulum protein Hsp47 binds specifically to the N-terminal globular domain of the amino-propeptide of the procollagen I alpha 1 (I)-chain. *J Cell Biochem*, 59:350-367.
232. Nagata K, Saga S and Yamada KM, (1988) Characterization of a novel transformation-sensitive heat-shock protein (HSP47) that binds to collagen. *Biochem Biophys Res Commun*, 153:428-434.
233. Nagata K, Hirayoshi K, Obara M, Saga S and Yamada KM, (1988) Biosynthesis of a novel transformation-sensitive heat-shock protein that binds to collagen. Regulation by mRNA levels and in vitro synthesis of a functional precursor. *J Biol Chem*, 263:8344-8349.
234. Hu G and Veis A, (1996) Posttranscriptional aspects of the biosynthesis of type 1 collagen pro-alpha chains: the effects of posttranslational modifications on synthesis pauses during elongation of the pro alpha 1 (I) chain. *J Cell Biochem*, 61:194-215.
235. Kivirikko KI and Pihlajaniemi T, (1998) Collagen hydroxylases and the protein disulfide isomerase subunit of prolyl 4-hydroxylases. *Adv Enzymol Relat Areas Mol Biol*, 72:325-398.
236. Peltonen L, Halila R and Ryhanen L, (1985) Enzymes converting procollagens to collagens. *J Cell Biochem*, 28:15-21.
237. Kessler E and Adar R, (1989) Type I procollagen C-proteinase from mouse fibroblasts. Purification and demonstration of a 55-kDa enhancer glycoprotein. *Eur J Biochem*, 186:115-121.
238. Imamura Y, Steiglitz BM and Greenspan DS, (1998) Bone morphogenetic protein-1 processes the NH2-terminal propeptide, and a furin-like proprotein convertase processes the COOH-terminal propeptide of pro-alpha1(V) collagen. *J Biol Chem*, 273:27511-27517.
239. Naughton BA, Dai Y, Sibanda B, Scharfmann R, San Roman J, Ziegler F and Verma IM, (1992) Long-term expression of a retrovirally introduced b-galactosidase gene in rodent cells implanted in vivo using biodegradable polymer meshes. *Somatic Cell and Molecular Genetics*, 18:451-462.
240. Krueger GG, Morgan JR, Jorgensen CM, Schmidt L, Li HL, Kwan MK, Boyce ST, Wiley HS, Kaplan J and Petersen MJ, (1994) Genetically modified skin to treat disease: potential and limitations. *J Invest Dermatol*, 103:76S-84S.

241. Krueger GG, Jorgensen CM, Petersen MJ, Mansbridge JN and Morgan JR, (1997) Use of clonally modified human fibroblasts to assess long-term survival *in vivo. Human Gene Therapy*, 8:523-532.
242. Krueger GG, Morgan JR and Petersen MJ, (1999) Biologic aspects of expression of stably integrated transgenes in cells of the skin in vitro and in vivo. *Proc Assoc Am Physicians*, 111:198-205.
243. Pierce GF and Mustoe TA, (1995) Pharmacologic enhancement of wound healing. *Annu Rev Med*, 46:467-481.
244. Guo N, Puhlev I, Brown DR, Mansbridge J and Levine F, (2000) Trehalose expression confers desiccation tolerance on human cells. *Nat Biotechnol*, 18:168-171.

Chapter 8

Adipose Tissue

Louise J Hutley, Felicity S Newell, Steven J Suchting and Johannes B Prins
Department of Diabetes and Endocrinology and University of Queensland Department of Medicine, Princess Alexandra Hospital, Ipswich Rd, Woolloongabba 4102, Australia. Tel: 0061–7–3240–2697; E-mail: jprins@medicine.pa.uq.edu.au

1. INTRODUCTION

Human adipose tissue metabolism is an important area of study, more so in recent times in view of the rising prevalence of obesity and its related disorders, including cardiovascular disease and diabetes. Understanding of adipose tissue metabolism in its broadest sense has been significantly promoted by the establishment of murine pre-adipocyte cell lines [1], and by robust systems for their culture and differentiation [2]. However, significant differences exist between these cell lines and primary human cells [3], and many reproducible and consistent findings obtained using the murine systems do not apply to human cells.

The relatively recent development of systems for the isolation, culture and differentiation of human pre-adipocytes [4-6], in addition to the more traditional systems for culture of human adipocytes and explants, has allowed many observations to be made that are more relevant to human physiology and pathology. However, as with many primary human cell cultures, the human system is more difficult, costly and time-consuming than the murine equivalents [3].

It is for these reasons that careful experimental design is paramount, as is an understanding of the type, quality and quantity of data that can be obtained. Consideration also needs to be given to human-specific characteristics that may influence experimental results or their interpretation. These include the great importance of the anatomical depot from which the adipose tissue/cells has been derived, and the understanding that the role(s) of leptin and the influence of brown adipose tissue (BAT) differ significantly between humans and rodents.

M.R. Koller, B.O. Palsson and J.R.W. Masters (eds.), Human Cell Culture Vol V, 173–187.

1.1 Organization of human adipose tissue

Adipose tissue is dynamic, with continuous turnover of stored triglyceride [7] and of adipose cells [8]. Net adipocyte triglyceride storage reflects the balance of lipolysis and lipogenesis (re-esterification); processes largely under hormonal control by insulin, corticosteroids and catecholamines. Adipocyte number reflects the balance of cell accumulation (by pre-adipocyte replication and differentiation) and cell deletion by apoptosis [9-12]. This latter feature has only recently been recognized, prompted by the discovery of pre-adipocytes and the demonstration of adipocyte apoptosis. It had previously been believed that there were only limited periods during life when acquisition of fat cells could occur, and that this process was irreversible [13, 14]. Therefore, adipose tissue mass reflects both the average size (determined predominantly by lipid content) and the number of adipose cells.

Pre-adipocytes are fibroblast-like cells localized to the stromo-vascular space of adipose tissue [15]. Identification and culture of these cells was first reported by Ng *et al.* in 1971 [16]. More recent work indicates that these cells can differentiate into adipocytes, myocytes or osteoblasts, supporting the concept of a common stem cell from which fat, muscle and bone cells are derived. Isolated pre-adipocytes are capable of limited replication *in vitro* [3], with a doubling time of approximately 60-70 hours (compared to 18-20 hours for human skin fibroblasts) (JB Prins, unpublished results). An *in vitro* system for complete differentiation of human pre-adipocytes was developed by Van *et al.* in 1978 [5], building upon earlier work reported by Poznanski *et al* [6]. The group of Ailhaud subsequently developed a serum-free system of pre-adipocyte differentiation in 1987 [4, 17, 18], which currently remains in use, with minor modifications. Human pre-adipocytes differentiate morphologically and accumulate intra-cytoplasmic lipid, and biochemically acquire protein and gene expression markers of mature adipocytes.

Adipocytes are distinguished morphologically by their large intra-cytoplasmic lipid droplet. They have energy storage, endocrine and possible immune roles as outlined below. Despite reports to the contrary [19, 20], it is generally accepted that adipocytes are fully differentiated and do not divide [21]. Thus, acquisition of adipose cells is generally believed to occur via pre-adipocyte replication and differentiation.

Adipose tissue loss occurs via reduction in intra-cytoplasmic lipid content (by lipolysis) and reduction in adipose cell number. The latter occurs via pre-adipocyte and adipocyte apoptosis, a process that is inducible *in vitro* by serum starvation [9, 12] or exposure to tumor necrosis factor (TNF) [12].

Significant anatomical depot-related differences in adipose tissue metabolism have been reported. These include differences in corticosteroid

[22] and catecholamine receptor number [23] and catecholamine effects [24, 25], gene expression patterns [26], differentiation capacity [27, 28] and susceptibility to apoptosis [11]. Many of these differences are likely to be clinically relevant, given the strong association between visceral obesity and metabolic disorders. Human adipose culture systems are a useful tool for exploring depot-specific characteristics of human adipose tissue.

1.2 Adipose tissue function

Traditionally, adipose tissue has been regarded as an energy storage depot and as an insulator. More recently, the role of the tissue has been expanded to include a major endocrine function and possibly a role in immunity.

In lean humans, there is approximately 100 times more stored energy in fat than in carbohydrate. The lipid is stored in a relatively water-free state, in contrast to carbohydrate which is heavily hydrated. The energy potential of fat is proportional to its saturation, and in mammals stored fat is maximally saturated, compatible with a liquid state. In lean individuals, most adipose tissue is stored subcutaneously. This affords considerable protection against heat loss and traumatic injury. The subcutaneous layer is not of uniform thickness, with marked gender and inter-individual differences. In females, more fat is distributed below the umbilicus, especially to the buttocks and thighs.

Adipocytes have autocrine, paracrine and endocrine functions [29, 30]. Inter-conversion of sex steroids occurs in adipose tissue, which is the major source of oestrogens in postmenopausal females and in males [15, 31]. Adipocytes and pre-adipocytes produce many growth factors, including insulin-like growth factor-1 (IGF-1), fibroblast growth factor (FGF), and transforming growth factors. Cytokines are also produced by adipose cells, and adipose cell-derived tumor necrosis factor-α is proposed to be a biochemical link between obesity and Type 2 diabetes [32]. The discovery by Friedman's group of leptin [33], an adipose-specific peptide hormone, confirmed the long-held hypothesis of a feedback system from adipose tissue to brain enabling the "sensing" of energy stores. It is now recognized that leptin levels regulate fertility in addition to appetite and the physiological response to low weight is a low leptin-mediated hypogonadotropic hypogonadism [34, 35]. Finally, adipocyte secretion of three of the rate-limiting components of the alternative complement pathway [36], Factors D (adipsin), B and C3, indicate a possible role in immunity. However, these factors are also integral to adipocyte production of acylation stimulating protein, suggesting the possibility of dual roles for the proteins [37].

2. TISSUE PROCUREMENT AND PROCESSING

2.1 General comments

Adipose tissue suitable for culture may be obtained at the time of surgery for an unrelated cause, or by either needle or open biopsy under local anesthetic. Visceral adipose tissue is only practically obtainable at laparotomy or laparoscopy, and availability is therefore limited to clinical centers. Tissue may be transported in either sterile saline-based solution such as Ringer's, or in culture medium (DMEM/Hams F12 or M-199) containing 10% fetal bovine serum (FBS). Small biopsies (less than 1cc) remain viable for long periods in serum-containing medium at 37°C, ultrastructural integrity is maintained for at least 14 days and pre-adipocyte isolation and culture has been successfully performed after 24 hours storage under these conditions. The methods outlined have been derived and modified from those described by Rodbell [38], Roncari [5], Hauner [4, 18, 27], Ailhaud [3] and Sugihara [20].

2.2 Isolation of pre-adipocytes and adipocytes

All tissue processing should be performed under as sterile conditions as are practicable.

1. The adipose tissue should be dissected free of visible nerves, blood vessels and fibrous tissue and then diced into 2-3 mm diameter pieces.
2. Add tissue to a collagenase digest solution (Hank's balanced salt solution (HBSS), 3 mg/ml Type 2 collagenase, 1.5% bovine serum albumin) and digested at 37°C for one hour. The ratio of digest solution to adipose tissue should be approximately 4:1.
3. Spin the resultant digest at 1500 *g* for 5 minutes to give a floating layer of mature adipocytes and a pre-adipocyte-containing pellet.
4. Gently draw off the adipocyte layer using a sterile transfer pipette and transfer to a sterile tube for further processing.
5. Discard the remaining supernatant.
6. Incubate the pellet in an erythrocyte lysis buffer (0.154 mmol/l NH_4Cl, 10 mmol/l $KHCO_3$, 0.1 mmol/l EDTA) for 10 minutes at room temperature.
7. Wash the remaining cells in phosphate-buffered saline (PBS, pH 7.4) three times.

3. CULTURE TECHNIQUES

3.1 Adipose tissue explants

Adipose tissue explants are readily cultured at 37°C under 5% CO_2 in serum-containing medium (eg DMEM/Ham's F12 or M199 containing 10% FBS) for long periods without loss of viability. We have demonstrated ultrastructural normality and are able to obtain good quality RNA from such explants after up to 14 days culture. Medium should be changed every 2-3 days.

3.2 Mature adipocytes

Human adipocytes are far less robust than their murine counterparts and as a result, culture is difficult and viability limited. The main difficulty arises from the buoyancy resulting from the high lipid content of the cells. This causes the cells to form a floating "cake" on the medium with sub-optimal culture conditions for much of the cell population. Gentle agitation or stirring can partially alleviate the problem, but the low resistance of the cells to shear stresses leads to cell lysis, manifest macroscopically as free lipid floating above the cells. Under these conditions, cell viability is limited to a few hours.

In an attempt to overcome this problem, the "ceiling culture" system has been developed by a number of groups including those of Ailhaud (personal communication) and Sugihara [20]. There are now many variations on the theme, but the basic principle is to allow the floating adipocytes to attach to a tissue culture-treated surface placed above them. Once attachment has occurred (about 24 hours) the surface can be inverted (to be right–side up) and culture continued as for monolayer cells such as pre-adipocytes (as outlined below). Practically, there is a choice of two methods. The first is to place the adipocytes into a T25 tissue culture flask, completely fill the flask with serum-containing medium and incubate the flask upside-down at 37°C for 24 hours to allow attachment. The second method is to inoculate a layer of adipocytes into a tissue culture dish containing serum-containing medium, and float a tissue culture treated coverslip(s) on the adipocyte layer. After 24 hr culture, the coverslips can be carefully removed, inverted, and placed into a petri dish and culture medium added.

An alternative solution is to culture the mature cells in a gel or matrix solution such as collagen, which allows adequate cellular nutrition. However, this system is expensive and difficult, and does not allow easy biochemical or histological assay of the cells, and therefore is little used.

3.3 Pre-adipocytes

The cell pellet resulting from the collagenase digest and erythrocyte lysis steps outlined above contains numerous cell types including pre-adipocytes, endothelial cells, fibroblasts and leukocytes. This cell pellet is cultured in serum-containing medium (as above) at 37°C under 5% CO_2 for 18-24 hr to allow attachment. The monolayer is then washed in PBS to remove non-attached cells, and fresh medium is added. Under these conditions, the majority of the attached cells are pre-adipocytes – fibroblast overgrowth is uncommon, presumably due to the low numbers of these cells in adipose tissue, and endothelial cells are more stringent than pre-adipocytes in their culture requirements and so rarely remain viable.

If a confluent initial culture is required, cells should be seeded to at least 20,000 cells per cm^2. If cells are initally plated at sub-confluence, medium changes (with serum-containing medium) should be undertaken every three days. Monolayers can be sub-cultured when 80-90% confluent, but will only tolerate a maximum of a 1:4 split. As outlined above, the growth rate is slow, so meticulous technique is essential. Even so, contamination (more commonly fungal) is not uncommon.

Pre-adipocytes are fibroblastic in morphology and display significant contact inhibition. Therefore, confluence in the true sense (100% coverage of the culture surface) is never achieved. Allowing cells to reach confluence during subculture significantly compromises the differentiation potential of the culture (discussed in more detail below). Therefore, the experimental aim should be borne in mind from an early stage.

3.4 Pre-adipocyte differentiation

Pre-adipocyte differentiation represents a co-ordinated process involving sequential alteration in patterns of gene expression, protein expression and function and morphological changes. Detailed understanding of the process in human cells is lacking, but much is known of the process in the murine system, and has been discussed in detail in reviews [39, 40].

There are a number of human-specific features of pre-adipocyte differentiation that warrant early discussion, as they greatly influence experimental design. Firstly, *in vitro* human pre-adipocyte differentiation only occurs efficiently in the absence of serum. As such, the development of a serum-free chemically modified differentiation medium represented a major advance in human adipose cell culture [17]. Secondly, the differentiation potential of human pre-adipocytes is inversely proportional to both the time spent in serum-containing medium and to the number of phases of replication allowed before differentiation [4]. In addition (as outlined above), allowing cells to reach confluence during subculture markedly limits differentiation potential. Thirdly, confluence of the pre-adipocyte monolayer just prior to differentiation greatly improves the efficiency of the process. Fourthly, anatomical depot-specific differences in differentiation potential exist, with visceral pre-adipocytes being refractory under most culture conditions [4, 28]. Finally, activators of the nuclear hormone receptor, peroxisome proliferator-activated receptor γ (PPARγ), promote differentiation of (particularly subcutaneous) pre-adipocytes [28].

The protocol for human pre-adipocyte differentiation is:

1. Ensure the pre-adipocyte monolayer is confluent. This is achieved either by calculating a sufficiently dense initial plating (as outlined in Section 3.3) or by allowing sufficient culture time in serum-containing medium for the cells to reach confluence. The latter consideration is important irrespective of previous subcultures.
2. Remove serum-containing medium and wash three times with sterile PBS (pH 7.4).
3. Add serum-free chemically modified medium (DMEM/Ham's F12, 15 mM $NaHCO_3$, 15 mM HEPES, 33 μM biotin, 17 μM pantothenate, 10 μg/ml human transferrin, 0.5 μM insulin, 0.2 nM tri-iodothyronine, 0.1 μM cortisol, 2 mM L-glutamine and antibiotics).
4. To promote differentiation, thiazolidinedione may be added to this medium (e.g. 0.1 μM Rosiglitazone) [28].
5. Culture at 37°C under 5% CO_2 in a humidified environment.
6. Change medium every 2-3 days.
7. Assess differentiation as outlined below, the process takes approximately 21 days.

4. ASSAY TECHNIQUES

Morphologically, pre-adipocytes (prior to differentiation) are indistinguishable from other human primary fibroblasts, even at the ultrastructural level. Similarly, there are no biochemical markers unique to pre-adipocytes in routine use. Therefore, one has to assume that the majority of cells isolated as outlined in section 2.2 are pre-adipocytes. In practice, this assumption is a reasonable one. Furthermore, this assumption is tested during the differentiation process when morphological features and biochemical markers are assessed.

4.1 Morphological features of human pre-adipocyte differentiation

Morphological changes in human pre-adipocytes are apparent within hours of changing the medium over the confluent monolayer to serum-free chemically modified differentiation medium. Cells "round up" and lose their fibroblastic morphology, and retain only sparse and tenuous cell-cell contact.

As differentiation progresses, intra-cytoplasmic lipid droplets form. These are easily seen under phase-contrast microscopy, and stain with Oil red-O or Nile red (see below). As differentiation continues, more droplets form such that the cytoplasm increases in size and becomes packed with small lipid droplets. Under these culture conditions, the lipid droplets do not coalesce to form a single droplet as is seen in mature adipocytes *in vivo*.

A feature of the human culture system for pre-adipocyte differentiation is its inefficiency. Under even the best of conditions, only 60-80% of cells will accumulate lipid (compared with close to 100% in murine 3T3-L1 cells). Cells tend to differentiate in clumps, with those cells that are refractory to differentiation reverting to a more fibroblastic morphology. As outlined above, a number of factors may influence the differentiation potential of a culture, but at least in subcutaneous cells, thiazolidinedione appears to reverse the detrimental affect of serum exposure and subculturing.

4.1.1 Assays of morphological changes and lipid accumulation

Phase contrast microscopy provides a simple means of assessing intra-cytoplasmic lipid accumulation, the classical morphological marker of differentiation. The use of an inverted microscope allows daily assessment of cultures and also provides a simple means of ensuring cultures are free of contamination. However, phase contrast microscopy is non-quantitative. Histological or fluorescent staining of intra-cytoplasmic triglyceride is straightforward and inexpensive, and may be used to quantitate lipid accumulation.

4.1.1.1 Oil Red-O staining

Oil Red-O is a stain soluble in neutral lipid, giving a red-orange color under white light [41]. It may be used alone or counter-staining may be added to demonstrate nuclear and/or cytoplasmic detail. The following method is quick and robust and stained monolayers may be stored for long periods to enable later analysis, photography or comparison. The assay may be adapted to small well sizes.

1. Remove culture medium and wash cells three times in phosphate buffered saline (PBS, pH 7.4).
2. Fix cells with 10% formalin in isotonic phosphate buffer for 1 hour at room temperature.
3. Immerse cells in Oil red-O working solution for 2 hours at room temperature. The stock solution is prepared by dissolving 0.74 g Oil red-O in 200 mL isopropanol and mixing overnight at 4°C on magnetic stirrer. Stock solution is filtered and stored at 4°C. Prior to use, the Oil red-O working solution is prepared by mixing stock solution with laboratory grade water at a ratio of 3 parts stock to 2 parts water, and leaving to stand at 4°C overnight, followed by a further filtration step.
4. Rinse thoroughly with water for 1-2 minutes (care must be taken as cells are easily washed off surface of culture dish).
5. Counter-stain with commercial hematoxylin for not more than 45 seconds.
6. Rinse gently with water (as above) for 5 minutes and allow to dry.
7. Apply drop of aqueous mounting medium and examine cells using light or phase contrast microscopy.

4.1.1.2 Nile Red staining

Nile Red is a fluorescent lipophilic dye that fluoresces with characteristic emmission spectra when in contact with triglyceride. Thus, it may be used in a fluorescence microscope to demonstrate lipid accumulation, like Oil red-O [42-44]. It can also be used in a suspension assay for lipid quantitation purposes, using a spectrofluorometer. The microscopic assay can be performed in 96-well plates, but the suspension assay requires a minimum of 10 cm^2 of cells (e.g. one well of a 6-well plate).

Microscopy assay

1. Remove culture medium and wash cells three times in PBS (pH 7.4).
2. Fix the cells at room temperature in 4% paraformaldehyde (in PBS, pH 7.4).
3. Wash three times with PBS.
4. Add PBS containing 1μg/ml Nile Red (stock solution is prepared in dimethyl sulfoxide (DMSO) to a final concentration of 1mg/ml). Mix and incubate at room temperature for 5-7 minutes.
5. Wash cells three times with PBS.
6. Examine cells under blue light (470 nm excitation). Lipid droplets will fluoresce bright yellow/gold.

Spectrofluorometric assay

1. Remove medium and wash cells three times in PBS (pH 7.4).
2. Add a minimal amount of trypsin-versene (e.g. 150 µl per well of a 6 well plate) to the cells. Incubate at 37°C for 10 minutes or until cells are detached from the culture plate.
3. Add PBS containing Nile Red to a final concentration of 1 µg/ml. Mix and incubate at room temperature for 5-7 minutes.
4. Measure yellow/gold fluorescence at room temperature in a spectrofluorometer at 488 nm excitation / 540 nm emission.
5. Normalize results to cell number or protein concentration where possible. In practice, we rarely obtain sufficient cells for these measurements when harvesting human pre-adipocytes from areas less than 25 cm^2. We therefore usually normalize to surface area or cell density (measured at OD_{600}).

4.1.2 Biochemical assays

Biochemical assays have the advantage of being functional. Lipoprotein lipase (LPL) is an enzyme whose expression occurs relatively early in pre-adipocyte differentiation [45]. Thus, it is widely used as an early marker. The assay itself, though widely reported in the earlier literature, is difficult to perform without wide intra- and inter-assay variability [46, 47]. It has therefore lost favor among most workers. Glycerol 3-phosphate dehydrogenase (G3PDH) is expressed much later in differentiation (JB Prins, unpublished results) and, along with adipsin and leptin, is a well-accepted late marker of differentiation. Leptin is a peptide hormone secreted by mature fat cells. We and others have been unable to identify intracellular leptin stores by either Western blotting of whole cell lysates or by immunofluorescence, suggesting that the peptide has a relatively rapid transit through the secretory pathway. Therefore, leptin secretion into the culture medium provides a ready means of assay. Thus, LPL or G3PDH activity or leptin production provide evidence of successful pre-adipocyte isolation and differentiation.

4.1.2.1 Glycerol-3-phosphate dehydrogenase activity

The enzyme glycerol 3-phosphate dehydrogenase produces α-glycerol phosphate and NAD^+ from dihydroxyacetone phosphate (DAP) and $NADH_2$ to provide the glycerol backbone for triacylglycerol formation. The conversion of NAD^+ to NADH may be measured spectrophotometrically at 340 nm. To obtain measurable results from human adipocytes, the use of culture areas of 25 cm^2 or above is recommended. Care should be taken to ensure samples are kept at 4°C at all times.

1. Wash cells twice with ice-cold PBS (pH 7.4).
2. Scrape cells into ice-cold harvest solution (50 mM Tris, pH 7.5, 1 mM EDTA and 500 μM DTT) and transfer to pre-chilled microcentrifuge tubes.
3. Disrupt cells by sonication and centrifuge lysates at 12,000 *g* for 15 minutes at 4°C.
4. Assay supernatant for G3PDH enzyme activity by combining in G3PDH assay mixture (final concentration 100 mM triethanolamine-HCl, pH 7.5, 2.5 mM EDTA, 50 μM DTT) and NADH (final concentration 0.24 mM) and bring to 37°C.
5. Initiate the reaction by adding DAP (final concentration 0.4 mM) and follow the A_{340} for at least four minutes to obtain an initial reaction rate.
6. Assay each supernatant in duplicate with a suitable reagent blank containing water instead of DAP.
7. Normalize the results to the protein content of each supernatant (which can be frozen and thawed).

Results are expressed as mU/mg supernatant protein, with 1 mU of enzyme activity being the amount catalyzing the oxidation of 1 nmol NADH per minute (calculated using the nM extinction coefficient of NADH at 340 nm as 6.22×10^{-3}).

4.1.2.2 Leptin secretion

Cells are cultured for 48 hours and the medium is then removed. Leptin levels are assayed using a commercially available human leptin assay kit (e.g. Linco radio-immunoassay). Dependent on the stage of differentiation and the sensitivity of the assay, monolayer areas of up to 25 cm^2 may be necessary for reproducible results.

4.1.3 Gene expression assays

Expression of genes encoding early or late markers of differentiation may be assessed by Northern blotting, protection assays or RT-PCR. Pre-adipocytes and adipose tissue (due to large cell size, low cell number per cm^2 and the high proportion of triglyceride, respectively) provide a low yield of RNA. In practical terms, a confluent T75 flask of differentiated human pre-adipocytes yields only 10-20 μg of total RNA, barely enough for one Northern blot. However, LPL, G3PDH and leptin mRNAs are highly expressed, so that stripping and re-probing the blot is possible. In contrast, RT-PCR only requires 0.5-1 μg RNA, but quantitative assays require considerably more than this.

Total RNA may be successfully extracted using the method of Chomczynski and Sacchi [48], or more recent commercial kit modifications (e.g. Tri-reagent or Trizol). The layer of free lipid must be pipetted off during the post-phenol/chloroform centrifugation step. Probe and primer sequences for numerous markers of differentiation are widely published.

5. UTILITY OF THE SYSTEM

Culture of human adipose cells enables the *in vitro* study of metabolic and growth characteristics of human adipose tissue(s) including comparisons of tissue from within and between individuals, and analysis of the effects of drugs/compounds on the above processes. Adipose characteristics amenable to *in vitro* study include:

- glucose uptake and insulin action
- lipogenesis
- lipolysis
- pre-adipocyte replication
- pre-adipocyte differentiation
- pre-adipocyte apoptosis
- adipocyte apoptosis
- gene expression
- protein and steroid production and metabolism

Unfortunately, the system is inefficient and dependent on access to human adipose tissue. However, despite these limitations, considerable evidence exists confirming that *in vitro* findings correlate well to *in vivo* adipose tissue function. This makes more than worthwhile the effort needed to overcome the inherent difficulties of the system.

We believe that human adipose culture systems remain under-utilized as a tool for investigation of adipose tissue biology and function, and for assessing the effects of drugs and other compounds.

6. SUMMARY

Successful culture of human adipose cells is fruitful and rewarding, enabling safe and non-invasive assessment of human adipose tissue function. Given the recognition of the central role of adipose tissue in many endocrine, metabolic and immunological systems, including appetite control and glucose and lipid metabolism, well-designed *in vitro* studies should provide considerable insight into both the physiology and the patho-physiology of the systems. The potential for direct study of the effects of known compounds on adipose tissue metabolism, as well as the potential for development of assay systems with utility for drug discovery, is largely untapped. The systems outlined above provide the opportunity to learn more about the normal physiology and biochemistry of human tissues.

REFERENCES

1. Green H and O Kehinde (1974) Sublines of mouse 3T3 cells that accumulate lipid. *Cell.* 1:113-116.
2. Smyth MJ, RL Sparks and W Wharton (1993) Proadipocyte cell lines: models of cellular proliferation and differentiation. *J. Cell. Sci.* 106:1-9.
3. Ailhaud G (1982) Adipose cell differentiation in culture. *Mol. Cell. Biochem.* 49:17-31.
4. Hauner H, G Entenmann, M Wabitsch, D Gaillard, G Ailhaud, R Negrel and EF Pfeiffer (1989) Promoting effect of glucocorticoids on the differentiation of human adipocyte precursor cells cultured in a chemically defined medium. *J. Clin. Invest.* 84:1663-1670.
5. Van RLR and AK Roncari (1978) Complete differentiation of adipocyte precursors. *Cell. Tiss. Res.* 195:317-329.
6. Poznanski WJ, I Waheed and R Van (1973) Human Fat Cell Precursors. *Lab. Invest.* 29:570-576.
7. Arner P (1988) Control of lipolysis and its relevance to development of obesity in man. *Diabetes Metab. Rev.* 4:507-515.
8. Prins JB and S O'Rahilly (1997) Regulation of adipose cell number in man. *Clin. Sci.* 92:3-11.
9. Prins JB, NI Walker, CM Winterford and DP Cameron (1994) Apoptosis of human adipocytes in vitro. *Biochem. Biophys. Res. Commun.* 201:500-507.
10. Prins JB, NI Walker, CM Winterford and DP Cameron (1994) Human adipocyte apoptosis occurs in malignancy. *Biochem. Biophys. Res. Commun.* 205:625-630.
11. Niesler CU, K Siddle and JB Prins (1998) Human pre-adipocytes display a depot-specific susceptibility to apoptosis. *Diabetes.* 47:1365-1368.
12. Prins JB, CU Niesler, CM Winterford, NA Bright, K Siddle, S O'Rahilly, NI Walker and DP Cameron (1997) Tumor necrosis factor-α induces apoptosis of human adipose cells. *Diabetes.* 46:1939-1944.
13. Ailhaud G (1990) Extracellular Factors, Signalling Pathways and Differentiation of Adipose Precursor Cells. *Current Opinion in Cell Biology.* 2:1043-1049.
14. Ailhaud G, *et al.* (1991) Growth and Differentiation of Regional Adipose Tissue: Molecular and Hormonal Mechanisms. *Int. J. Obes.* 15:87-90.
15. Ailhaud G, P Grimaldi and R Negrel (1992) A molecular view of adipose tissue. *Int. J. Obesity. Supp.* 16:s17-s21.
16. Ng CW, WJ Poznanski, M Borowieki and G Reimer (1971) Differences in growth in vitro of adipose cells from normal and obese patients. *Nature.* 231:445.
17. Deslex S, R Negrel, C Vannier, J Etienne and G Ailhaud (1986) Differentiation of Human Adipocyte Precursors in a Chemically Defined Serum-free Medium. *Int. J. Obes.* 10:19-27.
18. Hauner H, P Schmid and EF Pfeiffer (1987) Glucocorticoids and insulin promote the differentiation of human adipocyte precursor cells into fat cells. *J. Clin. Endocrinol. Metab.* 64:832-835.
19. Loffler G, G Mirter and R Hartrampf (1994) Culture of mature adipocytes: Evidence for cell division induced by growth factors. *Int. J. Obesity. Supp.* 18:88.

20. Sugihara H, N Yonemitsu, S Miyabara and K Yum (1986) Primary cultures of unilocular fat cells: Characteristics of growth in vitro and changes in differentiation properties. *Differentiation.* 31:42-49.
21. Negrel R (1994) Fat cells cannot divide. *Int. J. Obesity. Supp.* 18:88.
22. Rebuffe-Scrive M, M Bronnegard, A Nillson, J Eldh, J-A Gustafsson and P Bjorntorp (1990) Steroid Hormone Receptors in Human Adipose Tissue. *J. Clin. Endocrinol. Metab.* 71:1215-1219.
23. Hellmer J, C Marcus, T Sonnenfeld and P Arner (1992) Mechanisms for differences in lipolysis between human subcutaneous and omental fat cells. *J. Clin. Endocrinol. Metab.* 75:15-20.
24. Edens NK, SK Fried, JG Kral, J Hirsch and RL Leibel (1993) In vitro lipid synthesis in human adipose tissue from three abdominal sites. *Am. J. Physiol.* 265:E374-E379.
25. Rebuffe-Scrive M, B Andersson, L Olbe and P Bjorntorp (1989) Metabolism of Adipose Tissue in Intraabdominal Depots of Nonobese Men and Women. *Metab.* 38:453-458.
26. Montague CT, JP Prins, L Sanders, J Zhang, CP Sewter, JE Digby, CD Byrne and S O'Rahilly (1998) Depot-related gene expression in human subcutaneous and omental adipocytes. *Diabetes.* 47:1384-1391.
27. Hauner H, M Wabitsch and EF Pfeiffer (1988) Differentiation of Adipocyte Precursor Cells from Obese and Nonobese Women and From Different Adipose Tissue Sites. *Horm. Metab. Res. Supp.* 19:35-39.
28. Adams M, *et al.* (1997) Activators of PPARγ have depot-specific effects on human pre-adipocyte differentiation. *J. Clin. Invest.* 100:3149-3153.
29. Arner P (1994) Receptors in adipose tissue. *Int. J. Obesity. Supp.* 18 Supp 2:93.
30. Flier JS (1994) Adipose cells as endocrine cells. *Int. J. Obesity. Supp.* 18 Supp 2:93.
31. Boulton KL, DU Hudson, SW Coppack and KN Frayn (1992) Steroid hormone interconversions in human adipose tissue in vivo. *Metab.* 41:556-559.
32. Hotamisligil GS, NS Shargill and BM Spiegelman (1993) Adipose expression of tumor necrosis factor-α: Direct role in obesity-linked insulin resistance. *Science.* 259:87-91.
33. Zhang Y, R Proenca, M Maffei, M Barone, L Leopold and JM Friedman (1994) Positional cloning of the mouse obese gene and its human homologue. *Nature.* 372:425-432.
34. Flier JS (1998) What's in a name? In search of leptin's physiological role. *J. Clin. Endocrinol. Metab.* 83:1407-1413.
35. Ahima RS, D Prabakaran, C Mantzoros, D Qu, B Lowell, E Maratos-Flier and JS Flier (1996) Role of leptin in the neuroendocrine response to fasting. *Nature.* 382:250-252.
36. Choy LN, BS Rosen and BM Spiegelman (1992) Adipsin and an endogenous pathway of complement from adipose cells. *J. Biol. Chem.* 267:12736-12741.
37. Maslowska M, AD Sniderman, R Germinario and K Cianflone (1997) ASP stimulates glucose transport in cultured human adipocytes. *Int. J. Obes.* 21:261-266.
38. Rodbell M (1964) Metabolism of Isolated Fat Cells. 1. Effects of Hormones on Glucose Metabolism and Lipolysis. *J. Biol. Chem.* 239:375-380.
39. Mandrup S and MD Lane (1997) Regulating adipogenesis. *J. Biol. Chem.* 272:5367-5370.
40. Smas CM and HS Sul (1995) Control of adipocyte differentiation. *Biochem. J.* 309:697-710.
41. Ramirez-Zacarias JL, F Castro-Munozledo and W Kuri-Harcuch (1992) Quantitation of adipose conversion and triglycerides by staining intracytoplasmic lipids with oil red O. *Histochemistry.* 97:493-497.
42. Fowler SD and P Greenspan (1985) Application of Nile red, a fluorescent hydrophobic probe, for the detection of neutral lipid deposits in tissue sections: comparison with oil red O. *J Histochem Cytochem.* 33:833-6.

43. Greenspan P, EP Mayer and SD Fowler (1985) Nile red: A selective fluorescent stain for intracellular lipid droplets. *J. Cell. Biol.* 100:965-973.
44. Greenspan P, Fowler, S.D. (1985) Spectrofluorometric studies of the lipid probe, nile red. *J. Lipid Res.* 26:781-89.
45. Ailhaud G, P Grimaldi and J Negrel (1992) Cellular and Molecular Aspects of Adipose Tissue Development. *Annual Review of Nutrition.* 12:207-233.
46. Belfrage P and M Vaughan (1969) Simple liquid-liquid partition system for isolation of labeled oleic acid from mixtures with glycerides. *J. Lipid. Res.* 10:341-344.
47. Taskinen M-R, EA Nikkila, JK Huttunen and H Hilden (1980) A micromethod for assay of lipoprotein lipase activity in needle biopsy samples of human adipose tissue and skeletal muscle. *Clin. Chim. Acta.* 104:107-117.
48. Chomczynski P and N Sacchi (1987) Single-step method of RNA isolation by guanidinium thiocyanate-phenol-chloroform extraction. *Anal. Biochem.* 162:156-159.

Chapter 9

Mesenchymal Stem Cells

Mark F Pittenger, Gabriel Mbalaviele, Marcia Black, Joseph D Mosca and Daniel R Marshak.
Osiris Therapeutics Inc., 2001 Aliceanna Street, Baltimore, MD 21231.
E-mail: mpittinger@osiristx.com

1. INTRODUCTION

Many adult mammalian tissues maintain a healthy state by continuous renewal involving cell turnover. In response to trauma, disease or overuse, the body either repairs or regenerates the tissue.These two possibilities are distinguished in that *regeneration* results in new tissue that is indistinguishable from the original tissue in its structural organization, cellular content and function, whereas *repair* results in a high content of fibroblastic tissue, scar formation, limited structural organization and impaired function. Certain tissues, including skin, intestine, epithelium, and skeletal muscle have regenerative ability owing to resident progenitor cells. Other regenerating tissues, such as liver, have differentiated cells that retain the ability to de-differentiate and re-enter a proliferating growth phase before differentiating once again. Many types of blood cells originate from hematopoietic stem cells (HSCs) present in the sinusoids of bone marrow. In addition to progenitor cells resident in tissues, multipotent stem cells capable of connective tissue regeneration reside in bone marrow. The *in vitro* and *in vivo* study of these bone marrow-derived mesenchymal stem cells (MSCs) is important in developing a comprehensive understanding of the dynamic processes that occur in regenerating tissues and the roles that MSCs play.

The characterization of proliferative fibroblastic marrow cells with the potential to differentiate has been explored from multiple species including mouse [1-7], guinea pig [8, 9], rat [10-14], rabbit [15-20], dog [21-23], horse [24, 25] and man [26-35 and references therein], and several reviews have been published [49-55]. While many of these reports suggested the stem cell nature of the cells under study, the characterization was often incomplete.

M.R. Koller, B.O. Palsson and J.R.W. Masters (eds.), Human Cell Culture Vol V, 189–207.

Recently, we described the isolation and characterization of a homogeneous (approximately 99%) population of cells from human bone marrow and demonstrated the differentiation of these cells to the chondrogenic, osteogenic and adipogenic lineages [31]. We showed that cells grown from a single clonal cell could differentiate into these three lineages, and therefore described the cells as human mesenchymal stem cells (hMSCs). This chapter will focus on the isolation, culture and characterization of hMSCs from bone marrow. hMSCs are the best characterized of the mammalian species, and we have isolated these cells from marrow samples of more than 500 volunteer donors. The cultivation of hMSCs permits an enhanced understanding of this important progenitor cell for multiple tissue types and the development of new therapeutic approaches to tissue regeneration.

Bone marrow has been recognized as a source of progenitor cells for the hematopoietic system for many years, but the study of connective tissue precursors in marrow began in earnest with the culturing of attachment-dependent cells from bone marrow by Friedenstein, Owen and colleagues [8-10, 18, 19]. In a series of fundamental papers, these researchers described the fibroblastic cells present in marrow that could be isolated by allowing them to attach to the culture surface and rinsing away the unattached cells. The attached cells were spindle-shaped initially but soon took on a fibroblastic appearance. The cells had a range of proliferative capacity, many failing to form colonies. However, the cultures expanded extensively to provide cells for growth and transplantation studies. The cells were denoted as fibroblastic colony-forming cells (FCFC), and similar cells have been termed bone marrow stromal cells, colony-forming units-fibroblastic (CFU-F), osteogenic precursor cells (OPC), and marrow progenitor cells [8]. When cells were implanted in diffusion chambers, subsequent histology revealed that bone and cartilage were derived from the implanted cells [9]. This suggested the presence of either an osteochondral precursor cell or a mixture of osteocyte and chondrocyte precursors among the implanted cells. Many similar implant studies were subsequently done using this implantation scheme with progenitor cells derived from different sources [18, 19, 35, 40]. More recently, it has been possible to use genetically different syngeneic cells, or to transduce cells with foreign DNA as a marker to distinguish the infused or implanted cells from host cells [2, 3, 22, 37, 44]. These previous studies, however, did not demonstrate the multilineage potential of individual cells.

The isolation and culture of a homogeneous population of hMSCs from bone marrow was first accomplished by Caplan, Haynesworth and colleagues [26], and they prepared antibodies that proved useful in the characterization of the cells [41]. We have characterized and refined the isolation procedure to yield a reproducible and well-defined cellular preparation that is typically 99% homogeneous by flow cytometric measurements [31]. Using these highly purified hMSCs, it is possible to demonstrate their potential to progress to multiple mesenchymal lineages by *in vitro* assays. Moreover, there is no indication of multiple differentiated lineages in these assays, and the cells attain the desired mature cell type within 2 to 3 weeks. Here we describe in detail the isolation of hMSCs from bone marrow taken from the iliac crest, and their directed differentiation to the adipogenic, osteogenic and chondrogenic lineages *in vitro*. The hMSCs were never seen to differentiate spontaneously in culture to these lineages and, under the described conditions, nearly all of the hMSCs acquire the desired phenotype without the presence of other lineages.

Monoclonal antibodies developed against surface molecules [41] have proven useful to identify hMSCs during isolation and culture expansion, although no single antibody has been found that unequivocally identifies the hMSC. Antibodies to hMSCs also provide markers to follow during osteogenic differentiation of the cells [61]. Characterization of additional surface markers on hMSCs provides an extensive molecular profile to characterize these cells [31, 42, 43]. hMSCs can influence nearby or distant cell populations through the release of factors such as leukemia inhibitory factor (LIF), stem cell factor (SCF), granulocyte-colony-stimulating factor (G-CSF), interleukin-1 (IL-1), IL-6, and IL-11 [43, 44, 60].

A thorough discussion of the replicative potential of marrow-derived hMSCs has been published [27], demonstrating an average potential of 38±4 population doublings for hMSCs from adult donors. hMSCs have been isolated from marrow of all ages of donors, including one of 93 years (SE Haynesworth, personal communication). We estimate that bone marrow contains hMSCs at a frequency of $1{:}10^4$ to $1{:}10^5$, about 10 to 100-fold less abundant than hematopoietic stem cells. Given the ability of hMSCs to proliferate *in vitro*, a potentially therapeutic dose of 100 million hMSCs or more can be generated readily from a 20ml marrow aspirate from most individuals by passage two (see Table 2).

2. TISSUE PROCUREMENT AND PROCESSING

2.1 Safety from Biohazards - Universal Precautions

Researchers should be familiar with FDA Code of Federal Regulations (29 CFR Chapter XVII, 7-1-96 Edition), Occupational Safety and Health Administration (Labor § 1910.1030), and practice universal precautions when handling samples of human origin. The bone marrow donors are pre-tested for HIV and hepatitis, but tests are not available for all pathological agents. Personal protective equipment should include a disposable gown, double gloves, safety glasses or shield, and face mask.

2.2 Sources of hMSCs

Our donor population for bone marrow has routinely consisted of healthy male and female volunteers, 18-45 years old. Eligibility criteria include, but are not limited to, negative results for HIV, hepatitis B and hepatitis C, weight not greater than 10% above the normal body weight for height, no blood or bleeding disorders or abnormal scarring tendency, and no prescription medications other than those deemed allowable by the interviewing medical professional. There should be a minimum of 8 weeks between bone marrow donations and participation in other donor programs. The bone marrow donation consists of approximately 20-30ml of bone marrow withdrawn from the iliac crest (top of the pelvic bone). A local anesthetic is injected to the area, followed by bone marrow withdrawal using an aspiration needle by a physician or physician's assistant. The bone marrow samples are withdrawn bilaterally (10-15ml each side) from the back surface of each pelvic bone into syringes containing 3000 units of heparin. The actual bone marrow withdrawal takes about 5 to 15 minutes for each side. For cell isolation, bone marrow is withdrawn, thoroughly mixed with heparin, and may be held for up to 24 hours at room temperature before processing, without detrimental effects.

2.3 Marrow Processing

The bone marrow is washed by separating the total volume into 50ml centrifuge tubes at 5-7ml per tube and bringing the volume up to the 45ml mark with Dulbecco's modified phosphate buffered saline (DPBS). To remove fat, lysed cells and other undesired components, these tubes are centrifuged at 900*g* for 10 minutes at room temperature. The pellets are very soft. Carefully remove most of the supernatant and discard. Combine the pellets into one tube, gently mix, and perform an initial washed cell count. The average cells per ml from the washed bone marrow from a normal donor should be about 2-4 x 10^7. Once a cell concentration and a total cell number have been calculated, the cells are carefully loaded onto a density cushion of 1.073g/ml. Approximately 25ml of Percoll (or Ficoll or other density medium) solution in a 50ml disposable centrifuge tube is used with a maximum of 4 x 10^7 cells/ml and 2 x 10^8 cells/tube, total volume not to exceed 35ml. Centrifuge the density gradients with the samples at 1160*g* for 30 minutes at room temperature and decelerate slowly. Once centrifuged, remove the top layer of the tube(s), including the interface containing the nucleated cell fraction with hMSCs, without disturbing the pelleted RBCs. Place the nucleated cell fraction into a sterile centrifuge tube and wash the cells to remove the density gradient medium by adding two volumes of DPBS and inverting to gently mix. Then centrifuge the samples at 900*g* for 5 minutes at room temperature to pellet nucleated cells (using centrifuge brake). Remove and discard the supernatant, leaving only about 0.5ml on the cell pellet. The washed cells are resuspended in culture medium and another cell count performed. These cells are considered the light density cells. The average recovery of the light density cells from the density gradient is 28-30%. Only about 0.1% of the isolated nucleated cells in the light density fraction are actually MSCs and give rise to cell colonies. The light density cells are plated at approximately 160,000 cells/cm^2 with culture medium (about 0.2 ml/cm^2). Culture medium consists of Dulbecco's Modified Eagles Medium (DMEM with 1g/l glucose) with 10% fetal bovine serum (FBS) from selected lots [24], and cells are maintained at 37°C, 95% humidity and 5% CO_2. The use of selected FBS lots has been important as only some lots (approximately 30%) of FBS maintain the proliferative and multi-lineage potential of hMSCs. The need for serum selection may change as serum-free medium is developed.

These primary cultures will be kept in the same vessel for 12-16 days, usually 14 days, with media changes at days 2, 5, 8 and 11 (and 14 if necessary). At the first feeding, at day 2, the cell morphology should be observed. There will be a few cells that are attached and elongated. These appear as elongated, fibroblast-like cells and many, if not all, represent hMSCs. The MSCs are further enriched by selective expansion and medium changes that remove most of the non-attached cells such as those of the hematopoietic lineages. There will be many small round cells, settled and floating. These are described as phase bright cells and many will be erythrocytes. Many hematopoietic and differentiated mesenchymal cells require glucocorticoids and combinations of sera for a selective growth advantage. The MSCs appear to thrive in media lacking added glucocorticoids. Apoptotic or floating cells are removed with each subsequent medium change. No special effort is needed at this time to eliminate the contaminating cells other than standard medium changes. They will be eliminated in subsequent medium changes due to the selective nature of the medium.

Table 1. Isolation of hMSCs from bone marrow

Average amount of marrow aspirate	25 ml at ~28 x 10^6 nucleated cells/ml
Number of nucleated cells	Approximately 700 x 10^6
Number of nucleated cells after density gradient	Approximately 210 x 10^6 (~30%
Seed into T185 flasks at 162,000/cm^2	recovery), average 6-7 flasks
Number MSCs at end of P_0 (12-16 days growth)	12 to 35 x 10^6 cells
Seed new T185 flasks at 10^6/flask	(2-5 x 10^6 per 185 cm^2 flask)
Number of MSCs at end of P_1	48 to 245 x 10^6 cells
At 1 week, seed new flasks at 10^6/ flask	
Number MSCs at end of P_2	192 to 1700 x 10^6 cells
(about 1 week after seeding)	

The cultures are fed subsequently with fresh medium every 3 days and observations are made, which may include cell morphologies and rough percentages of the phase bright cells remaining. The presence of mitotic cells and expansion of the fibroblastic elongated cells to form colonies will be noted. At the day 11 feeding, a decision is made as to when the cells in the flask should be trypsinized and subcultured. The expanding cell colonies should be subcultured between days 12-16, before the colonies begin to merge.

3. CULTURE TECHNIQUES

3.1 hMSC Expansion in culture

The primary cultures are kept in the same vessel for 12-16 days (an average of 14 days) with periodic feedings every 3 days. When mitosis is still evident in the rapidly growing cells, and either colonies are beginning to merge or the cells are 75% confluent, it is time to trypsinize and subculture the cells. Observations are recorded and photographs can be taken as part of the record. The medium is aspirated and the attached cell layer is gently rinsed with DPBS to remove traces of serum that inhibits trypsin activity. The DPBS wash is aspirated and enough trypsin/EDTA solution (0.05% and 0.53mM, respectively) is added to the vessel to cover the cell layer. The vessel is left at room temperature for about 3 minutes until >90% of the cells are rounded and beginning to lift off. Once this is observed, DMEM with 10% FBS is added to the cell suspension at 1:1 (v/v) and gently mixed. The FBS contains enough endogenous protease inhibitors to prevent further action of the trypsin. The cells are gently washed from the surface, transferred into a sterile centrifuge tube, and centrifuged at ~900*g* for 5 minutes at room temperature.

After centrifugation, the supernatant is removed and discarded. The cell pellet is then resuspended in culture medium and a cell count is performed with the hemacytometer. The trypan blue exclusion method is used to determine the viability of the cell population (generally >95%). After trypsin/EDTA treatment to remove them from the culture vessel, hMSCs in suspension may be kept on ice for up to 4 hours in complete medium without detrimental effects. Once the total cell number has been calculated, the cells may be seeded into new vessels using the plating densities as described in Table 2. The minimum number of cells retrieved from primary cell culture is about $1.8 \times 10^4 /cm^2$.

Table 2. Cell densities and medium amounts for various cell culture vessels.

Vessel Size	Area (sq cm)	Primary seeding density	Passaged seeding density	Volume (ml)
12 well	3.8	6.2×10^5	2.1×10^4	1.5
6 well	9.6	1.6×10^6	5.0×10^4	2.0
35mm	9.6	1.6×10^6	5.0×10^4	2.0
100mm	55	9.0×10^6	3.0×10^5	10.0
T-25	25	4.0×10^6	1.3×10^5	5.0
T-80	80	1.3×10^7	4.3×10^5	15.0
T-185	185	3.0×10^7	1.0×10^6	35.0

The hMSCs from a primary culture that have been trypsinized and replated are denoted passage one (P_1) for that donor. After the primary culture is expanded, the doubling time of a passaged culture is more rapid, approximately every 48-72 hours. P_1 cells are cultured in the same vessel for 6-7 days with a medium change at day 3 or 4. Cells should be harvested when 80-90% confluent before they begin to form a dense layer. Cultures of hMSCs can be expanded for up to 8-10 passages without detectable senescence, such as slow growth or large broad flat cells appearing in the culture, although there is donor to donor variability. We routinely use the cultured hMSC at passages 1 to 5. We have analyzed the karyotype of several late passage hMSC cultures (passage 12), and the samples had a normal karyotype at 400 band resolution.

3.2 Freezing cultured hMSCs

The hMSCs can be stored frozen and survive the freezing and thawing process. After a cell count has been performed, the cell suspension may be placed in the centrifuge once more and centrifuged at 900*g* for 5 minutes at room temperature. The supernatant is discarded and the cells are quickly resuspended to a known concentration in freezing medium, 90% FBS and 10% DMSO. Aliquot cells to cryovials as soon as resuspended in freezing medium, as the DMSO can be cytotoxic. The vials are placed in a cell freezing container (Nalgene) with isopropanol (at room temperature) and placed at –80°C overnight. The vials are then transferred to a –150°C vapor phase nitrogen freezer. We have kept hMSCs at –150°C for 2 years or longer with greater than 95% viability.

4. ASSAY TECHNIQUES

The cultured mesenchymal stem cells can be analyzed for multipotentiality *in vitro* and *in vivo*, although no single assay can be used unequivocally to define an hMSC. We use a variety of *in vitro* assays to analyze the hMSCs as described below. The homogeneity of the cultures is tested by flow cytometry. A consistent set of antigens representing several classes of surface molecules are found on the hMSCs including growth factor receptors, cell adhesion molecules, integrins and cytokine receptors (see Table 3). We have not found a surface marker whose presence alone is sufficient to distinguish MSCs, but flow cytometric analysis of multiple cell surface molecules serves as an assessment of the homogeneity of the cultured cells. Routinely, the hMSCs are assayed for the presence of several MSC surface markers such as those defined by antibodies SH-2 [59] and SH-3 (American Type Culture Collection accession numbers HB 10743 and HB 10744 respectively), CD29 and CD44. The cultures are also tested for surface molecules found on HSCs or other cells of hematopoietic origin to show these cells are no longer present in the cultures. This is readily done using antibodies to CD14, CD34, or CD45, or others that are not found on cultured hMSCs [31].

4.1 Flow Cytometry

For flow cytometry, the cells are harvested routinely from the tissue culture flasks by treatment with 0.05% (w/v) trypsin and 25mM EDTA in PBS. The cells, in solution at a concentration of 0.5 x 10^6 cells/ml, are stained for 20 minutes using an empirically determined amount of each antibody, generally using 10 to 20ul. Labeled cells are washed thoroughly with two volumes of PBS and fixed in Flow Buffer (1% paraformaldehyde, 0.1% sodium azide, and 0.5% bovine serum albumin in PBS). We analyze labeled hMSCs on a FACS Caliber or FACS Vantage (Becton Dickinson) by collecting 10,000 events using the Cell Quest software program. By the end of P_1, the hMSCs are routinely 98% or more pure by flow cytometry for surface molecules.

Table 3. Flow Cytometry Characterization of Human Mesenchymal Stem Cells

PRESENT		ABSENT	
FACTOR RECEPTORS			
IFNγR	CDw119	EGFR-3	
TNFIR	CD120a	Fas ligand	CD95
TNFIIR	CD120b	T4	CD4
TGFβIR			
TGFβIIR			
bFGFR			
PDGFR	CD140a		
Transferrin	CD71		
INTEGRINS			
α1	CD49a	α4	CD49d
α2	CD49b	αL	CD11a
α3	CD49c	β2	CD18
α5	CD49e		
α6	CD49f		
αv	CD51		
β1	CD29		
β3	CD61		
β4	CD104		
HEMATOPOIETIC MARKERS			
		T4	CD4
		Mo2	CD14
			CD34
		Leukocyte Antigen	CD45
CYTOKINE RECEPTORS			
IL-1R	CD121a	IL-2R	CD25
IL-3Rα	CD123	CD9	
IL-4R	CDw124	Thy-1	CD90
IL-7R	CDw127		
IL-6R	CD126		
MATRIX MOLECULES			
ICAM-1	CD54	ICAM-3	CD50
ICAM-2	CD102	E-Selectin	CD62E
VCAM-1	CD106	P-Selectin	CD62P
L-Selectin	CD62L	PECAM-1	CD31
LFA-3	CD58	vW Factor	
ALCAM	CD166	Cadherin 5	
Hyaluronate Rec	CD44	Lewis^{x}	CD15
Endoglin	CD105		

4.2 *In vitro* Differentiation Assays

Methods have been developed for the *in vitro* differentiation of cultured hMSCs to adipogenic [31], chondrogenic [29-32] or osteogenic [26, 27, 28] lineages. We have compared the differentiation potential of hMSCs to strains of human fibroblasts in these differentiation assays, and the fibroblasts failed to differentiate [31]. These conditions, therefore, do not force differentiation of other mesenchymal cell types, but do allow hMSCs to express their inherent multi-lineage potential. The conditions described below do not produce mixtures of differentiated cell types, but rather they lead to a high proportion of the hMSCs achieving the desired phenotype. The presence of other cell types can be tested by RT-PCR for gene products of other lineages as described [31]. Whether the hMSCs cultured in each of our differentiation conditions produce secreted factors essential for lineage progression is unclear. While many classes of purified molecules are likely to have effects on hMSCs, such factors do not appear necessary as exogenous additives for either commitment or differentiation. Such factors may include members of the TGF-β superfamily, including growth and differentiation factors (GDFs), bone morphogenetic proteins (BMPs) and cartilage-derived morphogenetic proteins (CDMPs).

4.2.1 Adipogenic Differentiation

To induce adipogenic differentiation, hMSCs are cultured as monolayers in DMEM (1.0g/l glucose) containing 10% FBS and antibiotics and are allowed to become confluent [31]. The cells are cultured for 3-7 days past confluency, and then adipogenic induction medium is added. This medium (MDI+I medium) contains 1μM dexamethasone and 0.5mM methyl-isobutylxanthine, 10μg/ml insulin, 100μM indomethacin and 10% FBS in DMEM (4.5g/l glucose). The hMSCs are incubated in this medium for 48-72 hours at 37°C with 5% CO_2. The medium is then changed to adipogenic maintenance medium (AM medium) containing 10μg/ml insulin and 10% FBS in DMEM (4.5g/l glucose) for 24 hours. The hMSCs are then retreated with MDI (or MDI+I) for a second treatment. This is repeated once more for a total of three MDI+I treatments. The cultures are normally maintained in AM for an additional week to allow accumulation of lipid prior to fixation. AM medium is routinely changed every 3 days. Lipid vacuoles are first detectable microscopically within the cells within 48 hours of the first MDI treatment. These lipid vacuoles enlarge over time and become apparent through the light microscope. The cells can also be stained with the lipophilic dyes Oil Red-O or Nile Red. Nile Red has the advantage that it is fluorescent and the degree of differentiation and lipid accumulation can be quantitatively measured with a fluorescence plate reader [31]. Staining with Nile Red does not limit the ability to subsequently stain with Oil Red-O for histological visualization. We have maintained adipocytic hMSC cultures for up to three months, and the cells were maintained as mature adipocytes with a single large lipid vacuole that displaces the nucleus to the side.

4.2.2 Chondrogenic Differentiation

Chondrogenic differentiation of hMSCs is induced by placing 2.5 x 10^5 hMSCs into defined chondrogenic medium and subjecting them to gentle centrifugation (800*g* for 5 minutes) in a 15ml conical polypropylene tube, where they consolidate into a cell mass or pellet within 24 hours [16, 29-32]. Chondrogenic medium (CM) consists of high glucose (4.5g/l) DMEM supplemented with 6.25μg/ml insulin, 6.2μg/ml transferrin, 6.25μg/ml selenous acid, 5.33μg/ml linoleic acid, 1.25mg/ml bovine serum albumin (ITS+, Collaborative Research, Cambridge, MA), 0.1μM dexamethasone, 10ng/ml TGF-β3, 50μg/ml ascorbate 2-phosphate, 2mM pyruvate and antibiotics. TGF-β3 is prepared fresh from lyophilized powder, and CM in cultures is routinely replaced every third day. The chondrogenic cell pellets often increase in size 2-3 fold over 3 weeks. At harvest, the samples are fixed in 4% formaldehyde, then embedded in OCT freezing medium, and sections cut and analyzed. Extracellular matrix molecules such as type II collagen and aggrecan can be detected by immunohistochemical methods [30, 58]. Sections of chondrogenic samples can also be stained with Safranin O to detect the accumulation of proteoglycans [29, 30]. Little differentiation occurs during the first week in chondrogenic medium, although some gene products are detected by RT-PCR and some Safranin O staining may be evident. At 7-14 days, sections are partially positive for chondrogenic hMSCs, with Safranin O or collagen II antibodies, and staining then increases at later times.

4.2.3 Osteogenic Differentiation

To promote osteogenic differentiation of hMSCs, 3 x 10^4 cells are seeded into 35mm dishes in DMEM with 10% FBS to produce subconfluent, monolayer cultures [31, 39]. After 24 hours, this medium is replaced with the same medium containing osteogenic supplements (OS, consisting of 50μM ascorbate 2-phosphate, 10mM β-glycerol phosphate, and 100nM dexamethasone). Medium is replaced every three days. Increases in alkaline phosphatase activity are seen at 4 days, with activity peaking at 7-10 days and then slowly decreasing. Alkaline phosphatase activity is measured following the manufacturer's recommendations, and cells may be fixed, stained and photographed (kit #85, Sigma Chemical, St. Louis, MO). Calcium deposition is examined by the von Kossa stain and through a quantitative measurement of calcium deposition following the manufacturer's protocol (kit #587, Sigma Chemical).

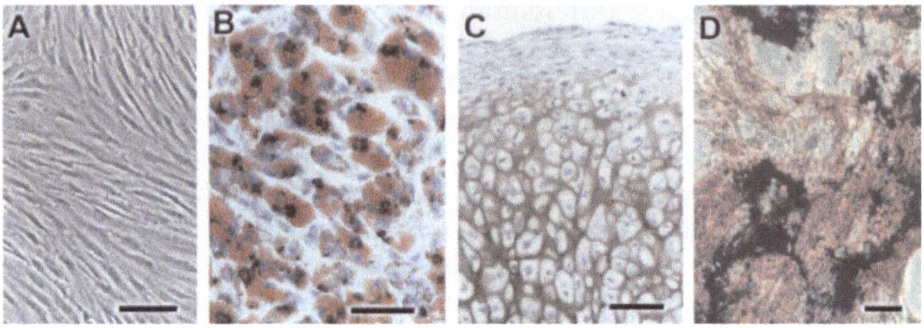

Figure 1. Differentiation of hMSCs to adipocytes, chondrocytes, and osteoblasts. A phase microscopy image of undifferentiated hMSCs is shown in panel A. Using the described conditions, the hMSCs will differentiate over 2-3 weeks to produce either adipocytes (B), chondrocytes (C) or osteoblasts (D). The adipocytes have been stained with oil red-O (which stains lipid vacuoles) and counter-stained with hematoxylin. The section of a cell pellet showing differentiated chondrocytes was stained with an antibody against type II collagen, and counter-stained with hematoxylin. The positive localization of type II collagen was revealed by the immunoperoxidase reaction product (brown). The nodules of aggregated osteoblastic cells have been stained for alkaline phosphatase (red) and mineral deposition (black: von Kossa silver stain). Bar equals 100 microns.

4.2.3 Stroma: support of hematopoiesis

The hMSCs in culture produce cytokines and hematopoietic growth factors, including IL-6, IL-7, IL-8, IL-11, IL-12, IL-14, LIF, SCF, *flt-3* ligand (FL) and macrophage-colony-stimulating factor (M-CSF) [42-44, 60]. To test hMSCs for their ability to support hematopoiesis *in vitro*, hMSCs are plated at 0.5 x 10^5 cells/cm^2 and cultured overnight at 37°C in 95% air and 5% CO_2. The hMSCs may be irradiated to prevent proliferation if desired, but it is not necessary as they become contact inhibited at confluency. For irradiation, hMSCs (1 x 10^6 cells/ml) are suspended in MSC medium and gamma irradiated with 1,600 rad. $CD34^+$ hematopoietic stem cells (HSC) are obtained from vendors or purified from healthy human bone marrow by immunoaffinity using an anti-CD34 antibody (Dynal). HSCs are resuspended at 5 x 10^4/ml in α-MEM-based medium containing 12.5% (v/v) horse serum, 12.5% (v/v) FBS, 0.2mM inositol, 20mM folic acid, 10^{-4}M 2-mercaptoethanol (MyeloCult 5100, Stem Cell Technologies), and plated onto monolayers of hMSCs. The cells are co-cultured for 5 weeks at a lower temperature of 33°C in 95% air and 5% CO_2 in the absence of added exogenous cytokines and growth factors. Every 3 days, half of the culture medium is gently removed to minimize loss of non-adherent cells, and replaced with an equal volume of fresh medium. At the end of the co-culture period, non-adherent HSC-derived cells are collected from the medium by centrifugation and counted. Any adherent HSC-derived cells are recovered by trypsin/EDTA treatment. Aliquots of HSC-derived cells are used for flow cytometry to determine the expression of hematopoietic lineage-specific surface markers. In addition, the cells from each fraction are tested for their colony-forming unit potential as follows.The cultured HSCs are resuspended at 0.5 x 10^5/ml, and 0.3ml of the suspension is added to 2.7ml of methylcellulose medium containing IMDM with 30% FBS, 1% bovine serum albumin, 100μM 2-mercaptoethanol, 2mM L-glutamine, 50ng/ml rhSCF, 20ng/ml granulocyte/macrophage-colony-stimulating factor (GM-CSF), 20ng/ml rhIL-3, 20ng/ml rhIL-6, 20ng/ml rhG-CSF, and 3U/ml rhEPO, (MyeloCult 4435, Stem Cell Technologies), and plated in duplicate into 35mm dishes. The HSCs are incubated at 37°C in 95% air and 5% CO_2. After 2 weeks, colonies derived from HSCs composed of >50 cells are scored, and the numbers from both fractions are combined for statistical analysis. The colonies will be of several types: BFU-E, CFU-F, CFU-GM and CFU-GEMM.

4.2.5 Early and Late Passage hMSCs

The assays described here are routinely performed using hMSCs at passage 1 to passage 5 and significant differentiation to each of the lineages is seen. At passages as late as 12 (approximately 30 billion-fold expansion), hMSCs underwent adipogenic differentiation similar to that seen at passage 4 (Beck and Pittenger, unpublished). A detailed study of hMSC osteogenic differentiation has been published [27], showing a gradual decline in response to OS by hMSCs beyond passage 8. The greatest decline in differentiation of hMSCs appeared to be the chondrogenic lineage, where beyond passage 5 chondrogenesis was not as robust as earlier passages. The abatement of chondrogenic potential was never accompanied by an increased propensity for one of the other lineages. One possibility is that the expansion culture conditions are not ideal for this lineage and that as hMSCs are expanded many-fold, they lose some lineage potential. Such loss of potential is not morphologically obvious nor discernible by flow cytometry. It is likely that further work with late passage hMSCs will improve the conditions necessary to maintain their multi-lineage potential at late passages. In additional studies, hMSCs retained their characteristic surface molecules through passage 12, when analyzed by flow cytometry.

4.3 *In vivo* Differentiation Assays

hMSCs can be tested for their ability to differentiate to lineages by implanting the cells into host animals. We have used the formation of bone and cartilage by hMSCs seeded onto ceramic cubes and implanted subcutaneously into athymic mice to assay FBS lots to be used for hMSC culture [46]. The potential of the cells to form bone and cartilage when placed in diffusion chambers and implanted in ectopic sites has also been useful in demonstrating some characteristics of hMSCs [18, 19, 35, 40]. The osteogenic potential of hMSCs at a site of bone repair in vivo has been evaluated in athymic rats [14], and a study of autologous MSC regeneration of a bone resection in a canine model [23] has demonstrated the potential of MSCs for therapeutic use. Similarly, results have been presented on the implantation of MSCs into the defects of articular cartilage of animal models [17], and this approach holds much therapeutic promise.

5. UTILITY OF THE SYSTEM

MSCs have enormous potential to aid the research community. MSCs have the advantage of extensive proliferation and expansion in culture, unlike many primary differentiated cells. The limitation of hMSCs is that they are not immortal and therefore become senescent at late passage. It is necessary, consequently, to re-isolate new primary cultures of hMSCs routinely. The availability of a commercial source of hMSCs (Clonetics/ Poietics Div. of BioWhittaker, Walkersville, MD) for research use will facilitate this process. However, researchers should replicate experiments on independent isolates of MSCs from multiple donors. Results should be validated for MSCs derived from both genders of different ages, and any specific findings related to age or gender will be of considerable interest. Age and gender are important parameters in considering models for certain degenerative diseases, such as osteoporosis.

The hMSCs are normal, diploid human cells, which are not transformed by oncogenes or viruses, as are many cell lines and other model systems. This feature permits hMSCs to be used to study signal transduction and gene regulation without the limits imposed by transformed cell lines. For example, growth control and transcriptional activation can be studied without the presence of SV40 large T antigen, adenovirus E1A, activated

Ras, or other immortalizing protein. While this is true of other primary cells in culture, they often show limited ability to grow and divide.

Screening of compounds using hMSCs can contribute to pharmaceutical development. The multiple pathways of hMSC differentiation offer the opportunity to identify genes important in these pathways and compounds that modulate such genes or their products. New activities of compounds might be discovered in an unexpected pathway. For example, a compound developed for bone metabolism could be linked to cartilage, fat, or other tissue types early in the research and development process. hMSCs also represent a convenient and rapid way to evaluate the toxicity and potency of compounds on normal human cells, and the ability of the hMSCs to differentiate to multiple lineages allows different tissue types to be evaluated. Unexpected side effects of new drug candidates may be avoided prior to launching full-scale toxicity testing in animals.

Functional genomics has received much attention recently as a means to produce new therapeutic targets for drug discovery. hMSCs can assist researchers in this effort, as the role of genes identified through genomics and bioinformatics can be tested quickly and easily, for safety and efficacy, in a particular functional differentiation pathway. This is an extraordinary advantage of time and cost over transgenic, knock-out, or knock-in mice to evaluate phenotypic changes. Rapid screening of gene activities using the hMSCs and their mesenchymal lineages may provide new insights into the function of genes within desired areas of development.

6. CONCLUDING REMARKS

While much remains to be learned about the role of hMSCs in tissue regeneration, their ability to grow clonally and to differentiate into multiple cell types *in vitro* establishes their stem cell nature. Basal nutrients, cell density, spatial organization and mechanical forces, as well as growth factors and cytokines, appear to have profound influences on differentiation of hMSCs. Upon initial isolation, the hMSCs from bone marrow are not in log phase growth and establish detectable colonies slowly. Whether this is due to the crisis of dealing with a new environment, a lack of necessary growth factors, cell-matrix or cell-cell interactions is unclear. For instance, as the hMSCs proliferate in culture, there is a concurrent loss of cells of the hematopoietic lineage, such as macrophages, monocytes, platelets, erythrocytes and HSCs. Do these cells provide factors to hMSCs? The cultivation of hMSCs will allow an enhanced understanding of this important progenitor of multiple tissue types, and the possibility of new therapeutic modalities based on *in vitro* expanded autologous cells. The ability of hMSCs to respond to appropriate signals is certain to be important for commitment, proper functional differentiation and integration with surrounding tissues [62]. Until detailed *in vivo* cell characterization is available, there will always remain some question as to whether the hMSCs isolated following the *in vitro* proliferative growth phase are identical to those found in undisturbed marrow. Nevertheless, the hMSCs isolated from marrow represent a renewable source of stem cells for multiple mesenchymal lineages, and much insight will be gained as research on them continues. By understanding the relationship between *in vitro* assays and *in vivo* implantation results, we may hope to understand the biology of these adult mesenchymal stem cells and develop therapeutic uses for marrow-derived human MSCs.

REFERENCES

1. Saito T, Dennis JE, Lennon DP, Young RG, and Caplan AI (1995) Myogenic expresssion of mesenchymal stem cells within myotubes of mdx mice in vitro and in vivo. *Tissue Engin.* 1(4): 327-343.
2. Pereira RF, Halford KW, O'Hara MD, Leeper DB, Sokolov BP, Pollard MD, Bagasra O, and Prockop DJ (1995) Cultured adherent cells from marrow can serve as long-lasting precursor cells for bone, cartilage and lung in irradiated mice. *Proc. Natl. Acad. Sci. USA.* 92:4857-4861.
3. Pereira RF, O'Hara MD, Laptev AV, Halford KW, Pollard MD, Class R, Simon D Livezey K and Prockop DJ (1998) Marrow stromal cells as a source of progenitor cells for nonhematopoietic tissues in transgenic mice with a phenotype of osteogenesis imperfecta. *Proc. Natl. Acad. Sci. USA* 95:1142-1147.
4. Cui Q, Wang GJ and Balian G (1997) Steroid-induced adipogenesis in a pluripotent cell line from bone marrow. *J. Bone Joint Surg.* 79:1054-1063.
5. Poliard A, Nifuji A, Lamblin D Forest C and Kellermann O (1995) Controlled conversion of an immortalized mesodermal progenitor cell towards osteogenic chondrogenic and adipogenic pathways. *J. Cell Biol.* 130:1461-1472.
6. Dennis JE, Merriam A, Awadalla A, Yoo JU, Johnstone B, and Caplan AI. (1999) A quadripotent mesenchymal progenitor cell isolated from the marrow of an adult mouse. *J. Bone Miner. Res.* 14:700-709.
7. Phinney DG, Kopen G, Isaacson RL, and Prockop D (1999) Plastic adherent stromal cells from the bone marrow of commonly used strains of inbred mice: variations in yield, growth and differentiation. *J. Cell. Biochem.* 72:570-585.
8. Friedenstein AJ (1976) Precursor cells of mechanocytes. *Int. Rev. Cytol.* 47:327-355.
9. Friedenstein AJ, Petrakova KV, Kurolesova AI, and Frolova GP (1968) Heterotopic transplants of bone marrow: Analysis of precursor cells for osteogenic and haematopoietic tissues. *Transplantation* 6:230-47.
10. Owen M (1970) The origin of bone cells. *Int. Rev. Cytol.* 28:213-238.
11. Grigoriadis AE, Heersche J, and Aubin JE (1990) Continuously growing bipotential and monopotential myogenic, adipogenic and chondrogenic subclones isolated from the multipotential RCJ3.1 clonal cell line. *Developmental Biology* 142:313-318.

12. LeBoy PS, Beresford J, Devlin C and Owen M (1991) Dexamethasone induction of osteoblast mRNAs in rat marrow stromal cell cultures. *J. Cell Physiol.* 146:370-378.
13. Beresford JN, Bennett JH, Devlin C, LeBoy PS, and Owen M (1992) Evidence for an inverse relationship between the differentiation of adipocytic and osteogenic cells in rat marrow stromal cell cultures. *J. Cell Sci.* 102:341-351.
14. Kadiyala S, Jaiswal N, and Bruder SP (1997) Culture-expanded bone marrow-derived mesenchynal stem cells can regenerate a critical-sized segmental bone defect. *Tissue. Engin.* 3:173-185.
15. Young RG, Butler DL, Weber W, Caplan AI, Gordon SL, and Fink DJ (1998) Use of mesenchymal stem cells in a collagen matrix for achilles tendon repair. *J. Orthop. Res.* 16: 406-413.
16. Johnstone B, Hering TM, Caplan AI, Goldberg VM, and Yoo J (1998) In vitro chondrogenesis of bone marrow-derived mesenchymal progenitor cells. *Exp. Cell Res.* 238:265-272.
17. Wakitani S, Goto T, Pineda SJ, Young RG, Mansour JM, Goldberg VM, and Caplan AI (1994) Mesenchymal cell based repair of large, full thickness defects of articular cartilage. *J. Bone Joint Surg.* 76:579-592.
18. Ashton BA, Allen TD, Howlett CR, Eaglesom CC, Hattori A and Owen M. (1980). Formation of bone and cartilage by marrow stromal cells in diffusion chambers *in vivo. Clin. Orth. Rel. Res.* 151:294-307.
19. Friedenstein AJ, Chailakhyan RK, Gerasimov UV (1987) Bone marrow osteogenic stem cells: In vitro cultivation and transplantation in diffusion chambers. *Cell Tissue Kinet.* 20:263-272.
20. Grande DA, Southerland SS, Ryhanna Manji BS, Pate DW, Schwartz RE, and Lucas PA (1995) Repair of articular defects using mesenchymal stem cells. *Tissue Engin.* 1:345-353.
21. Kadiyala S, Young RG, Thiede MA, and Bruder SP (1997) Culture expanded canine mesenchymal stem cells possess osteochondrogenic potential *in vivo* and *in vitro*. *Cell Transplantation* 6:125-134.
22. Hurwitz DR, Kirchgesser M, Merrill W, Galanopoulos T, McGrath CA, Emani S, Hansen M, Cherington V, Appel J, Bizinkauskas CB, Brackmann HH, Levine PH, and Greenberger JS (1997) Systemic delivery of human growth hormone or human factor IX in dogs by reintroduced genetically modified autologous bone marrow stromal cells. Hum. Gene Ther. 8:137-156.
23. Bruder SP, Kraus KH, Goldberg VM, and Kadiyala S (1998) The effect of autologous mesenchymal stem cell implants on the healing of canine segmental bone defects. J. Bone Joint Surg.80A: 985-996.
24. Fortier LA, Cable CS and Nixon AJ (1996) Chondrogenic differentiation of mesenchymal cells in long term three-dimensional fibrin cultures treated with IGF-1. *Trans. Orthop. Res. Soc.* 21:110.
25. Richardson DW, Univ. Pennsylvania School of Veterinary Medicine – personal communication.
26. Haynesworth SE, Goshima J, Goldberg, VM, and Caplan AI (1992) Characterization of cells with osteogenic potential from human bone marrow. *Bone. 13*:81-88.
27. Bruder SP, Jaiswal N and Haynesworth SE (1997) Growth kinetics, self-renewal and the osteogenic potential of purified human mesenchymal stem cells during extensive subcultivation and following cryopreservation. *Journal of Cellular Biochem.* 64:278-294.
28. Owen ME, Cave J, Joyner CJ (1987) Clonal analysis in vitro of osteogenic differentiation of marrow CFU-F. *J. Cell Sci.* 87:731-738.

29. Barry FP, Johnstone B, Pittenger MF, Mackay AM and Murphy JM (1997) Modulation of the chondrogenic potential of human bone marrow-derived mesenchymal stem cells by TGFβ1 and TGFβ3. *Trans. Ortho Res. Soc.* 22:228.
30. Mackay AM, Beck SC, Murphy JM, Barry FP, Chichester CO and Pittenger MF (1998) Chondrogenic differentiation of cultured human mesenchymal stem cells from marrow. *Tissue Engin.* 4:415-428.
31. Pittenger MF, Mackay AM, Beck SC, Jaiswal RK, Douglas R, Mosca JD, Moorman MA, Simonetti DW, Craig S, Marshak DR (1999) Multilineage potential of adult human mesenchymal stem cells. *Science* 284:143-147.
32. Yoo JU, Barthel TS, Nishimura K, Solchaga L, Caplan AI, Goldberg VM and Johnstone B (1998) The chondrogenic potential of human bone-marrow derived mesenchymal progenitor cells. *J. Bone Joint Surg.* 80A:1745-1757.
33. Cheng S-L, Yang JW, Rifas L, Zhang S-F, and Avioli LV (1994) Differentiation of human bone marrow osteogenic stromal cells in vitro: Induction of the osteoblastic phenotype by dexamethasone. *Endocrinology* 134:277-286.
34. Gronthos S, Ohta S, Graves SE, and Simmons PJ (1994) The STRO-1+ fraction of adult human bone marrow contains the osteogenic precursors. *Blood* 84:4164-4173.
35. Gundle R, Joyner C, Triffit J (1995) Human bone tissue formation in diffusion chamber culture in vivo by bone derived cells and marrow stromal fibroblastic cells. *Bone* 16:597-601.
36. Gimble JM, Morgan C, Kelly K, Wu X, Dandapani V, Wang CS and Rosen V (1995) Bone morphogenetic proteins inhibit adipocyte differentiation by bone marrow stromal cells. *J. Cell Biochem.* 58: 393-402.
37. Houghton A, Oyajobi BO, Foster GA, Russell RG and Stringer BMJ (1998) Immortalization of human marrow stromal cells by retroviral transduction with a temperature sensitive oncogene: Identification of bipotential precursor cells capable of directed differentiation to either an osteoblast or adipocyte phenotype. *Bone*. 22: 7-16.
38. Nuttall, ME, Patton AJ, Olivera DL, Nadeau DP, and Gowen M (1998) Human trabecular bone cells are able to express both osteoblastic and adipocytic phenotype: Implications for osteopenic disorders. *J. Bone and Mineral Res.* 13:371-382.
39. Jaiswal N, Haynesworth SE, Caplan AI, and Bruder SP (1997) Osteogenic differentiation of purified, culture-expanded human mesenchymal stem cells in vitro. *J. Cell. Biochem.* 64: 295-312.
40. Kuznetsov SA, Krebsbach PH, Satomura K, Kerr J, Riminucci M, Benayahu D, and Robey PG (1997) Single-colony derived strains of human marrow stromal fibroblasts form bone after transplantation in vitro. *J. Bone Miner. Res.* 12:1335-1347.
41. Haynesworth SE, Baber MA and Caplan AI (1992) Cell surface antigens on human marrow-derived mesenchymal cells are detected by monoclonal antibodies. *Bone* 13:69-80.
42. Majumdar M, Thiede MA, Mosca JD, Moorman M, and Gerson SL (1998) Phenotypic and functional comparison of cultures of marrow-derived mesenchymal stem cells (MSCs) and stromal cells. *J. Cell. Physiol.* 176:57-66.
43. Haynesworth SE, Baber MA, and Caplan AI (1996) Cytokine expression by human marrow-derived mesenchymal progenitor cells in vitro: effects of dexamethasone and IL-1α. *J. Cell. Physiol.* 166: 585-592.
44. Allay JA, Dennis JE, Haynesworth SE, Majumdar MK, Clapp DW, Schultz LD, Caplan AI (1997) LacZ and interleukin-3 expression in vivo after retroviral transduction of marrow-derived human osteogenic mesenchymal progenitors. *Hum Gene Therapy* 8:1417-1427.

45. Lazarus HM, Haynesworth SE, Gerson SL, Rosenthal NS, and Caplan AI (1995) Ex vivo expansion and subsequent infusion of human bone marrow-derived stromal progenitor cells (mesenchymal progenitor cells): Implications for therapeutic use. *Bone Marrow Transpl.* 16:557-564.
46. Lennon DP, Haynesworth SE, Bruder SP, Jaiswal N, and Caplan, AI (1996) Development of a Serum Screen for Mesenchymal Progenitor Cells from Bone Marrow. *In Vitro Animal Cellular & Developmental Biology* 32: 602-611.
47. Friedenstein, A J (1995) Marrow Stromal Fibroblasts. *Calcif. Tissue Int.* 56 Suppl1: S17.
48. Bennett, JH, Joyner CJ, Triffitt JT, Owen ME (1991) Adipocytic cells cultured from marrow have osteogenic potential. *J. Cell Science* 99:131-139.
49. Owen, ME, and Friedenstein AJ (1988) Stromal stem cells: marrow derived osteogenic precursors. *Ciba Found. Symp.* 136:42-60.
50. Owen M (1985) Lineage of osteogenic cells and their relationship to the stromal system. In *Bone and Mineral/3* WA Peck, ed. Amsterdam: Elsevier, pp1-25.
51. Beresford JN (1989) Osteogenic stem cells and the stromal system of bone and marrow. *Clin. Orthop.* 240:270-280.
52. Caplan, AI (1991) Mesenchymal stem cells. *J. Orthop. Res.* 9:641-650.
53. Bruder, SP, Fink DJ, Caplan AI (1994) Mesenchymal stem cells in bone development, bone repair and skeletal regeneration therapy. *J. Cellular Biochemistry* 56:283-294.
54. Caplan, AI (1994) The mesengenic process. *Clinics in Plastic Surgery* 21:429-435.
55. Prockop DJ (1998) Marrow stromal cells as stem cells for non-hematopoietic tissues. Science 276:71-74.
56. Bruder, SP, Kurth AA, Shea M, Hayes WC, Jaiswal N, Kadiyala S (1998) Bone regeneration by implantation of purified, culture-expanded human mesenchymal stem cells. *J. Ortho. Res.* 16:155-162.
57. Ballock, RT and Reddi, AH (1994) Thyroxine is the serum factor that regulates morphogenesis of columnar cartilage from isolated chondrocytes in chemically defined medium. *J. Cell Biol.* 119:1311-1317.
58. Srinivas, GR, Barrach, HJ and Chichester, CO (1993) Quantitative immunoassays for type II collagen and its cyanogen bromide peptides. *J. Immunol. Methods* 159:53-59.
59. Barry FP, Boynton RE, Haynesworth SE, Murphy JM, and Zaia J (1999) The monoclonal antibody SH-2, raised against human mesenchymal stem cells, recognizes an epitope on endoglin (CD105). Biochem. Biophys. Res. Commun. 265:134-139.
60. Mbalaviela G, Abu-Amer Y, Meng A, Jaiswal R, Beck S, Pittenger M, Thiede M and Marshak D. (2000) Activation of peroxisome proliferator-activated receptor γ pathway inhibits osteoclast differentiation. *J. Biol. Chem.* 275:14388–14393.
61. Bruder SP, Ricalton NS, Boynton RE, Connolly TJ, Jaiswal N, Zaia J and Barry FP (1998) Mesenchymal stem cell surface antigen SB-10 corresponds to activated leukocyte cell adhesion molecule and is involved in osteogenic differentiation. J. Bone Mineral Res. 13:655-663.
62. Jaiswal RK, Jaiswal N, Bruder SP, Mbalaviele G, Marshak D and Pittenger MF (2000) Adult human mesenchymal stem cell differentiation to the osteogenic or adipogenic lineage is regulated by mitogen-activated protein kinase. *J. Biol. Chem.* 275:9645–9652.

Chapter 10

Peripheral Blood Fibrocytes

Jason Chesney and Richard Bucala
The Picower Institute for Medical Research, Laboratory of Medical Biochemistry, 350 Community Drive, Manhasset, NY 11030. Tel: 001-516-562-9406; Fax: 001-516-365-5090; e-mail: rbucala@picower.edu

1. INTRODUCTION

Connective tissue fibroblasts are a quiescent cell population that under normal circumstances remain sparsely-distributed throughout the extracellular matrix [1]. As a consequence of injury, fibroblasts enter and proliferate within the injured site [2]. The precise origin of the fibroblast-like cells within wounds has been controversial since the original microscopic studies of developing connective tissue performed by Paget in 1863 [3, 4]. That wound fibroblasts appeared by migration from adjacent tissue was supported by experiments showing the apparent ingrowth of fibroblasts from local areas, and by the observation that India ink-tagged monocytes failed to develop into tissue fibroblasts *in vivo* [4, 5]. Other studies, however, reported evidence for the differentiation of leukocytes into fibroblasts within subcutaneous diffusion chambers, and the apparent *in vitro* transformation of peripheral blood mononuclear cells into collagen-producing cells [6, 7].

Several years ago, investigations into the cell population present in experimentally-implanted subcutaneous wound chambers led to the discovery of an adherent, proliferating cell type that displayed fibroblast properties yet expressed distinct hematopoietic/leukocyte cell surface markers [8]. Wound chambers consist of short lengths of sponge-filled, silastic tubing and are a frequently employed model for the study of tissue reparative responses *in vivo.* Implantation of these chambers into the subcutaneous space of mice results in a rapid infiltration of peripheral blood inflammatory cells, including neutrophils, monocytes, and lymphocytes [9, 10]. Large numbers of adherent, spindle-shaped cells that resemble fibroblasts were unexpectedly observed to infiltrate wound chambers soon after implantation, and coincidentally with the appearance of circulating

M.R. Koller, B.O. Palsson and J.R.W. Masters (eds.), Human Cell Culture Vol V, 209–219.

inflammatory cells. Double immunofluorescence studies showed that within 24 hours of implantation, as many as 10-15% of the cells present in wound chamber fluid stain positively both for Type I collagen and CD34 [8]. At the time, the cell surface marker CD34 was considered to be expressed exclusively by hematopoietic stem cells, and the combination of CD34 and collagen had not previously been described in any cell type. These studies suggested the presence in wounds of a previously uncharacterized cell type displaying fibroblast-like features, but expressing markers for bone marrow-derived cells. Follow-up immunohistochemical analysis of wound chambers that had been implanted in mice confirmed the presence of $CD34^+$ spindle-shaped cells in areas of collagen matrix deposition. These cells also were identified to be present in areas of scarring [8].

More detailed investigations have since demonstrated that peripheral blood fibrocytes comprise 0.1-0.5% of circulating leukocytes (J. Chesney, *unpublished observations*). Fibrocytes obtained from blood constitutively express in culture the fibroblast products collagen I, collagen III, and fibronectin, as well as the leukocyte common antigen CD45RO, the pan-myeloid antigen CD13 and the hematopoietic stem cell antigen CD34 [8]. Fibrocytes do not synthesize epithelial (cytokeratin), endothelial (von Willebrand factor VIII-related protein), or smooth muscle (α-actin) cell markers, and are negative for non-specific esterases as well as the monocyte/macrophage-specific marker, CD16 [8]. Fibrocytes also do not express proteins produced by dendritic cells or their precursors (CD25, CD10, and CD38) or the pan-B cell antigen CD19 [8, 11-13]. Scanning electron microscopy has shown these cells to be morphologically distinct from blood-borne leukocytes, and to display unique cytoplasmic extensions intermediate in size between microvilli and pseudopodia (Figure 1). Of note, studies employing sex-mismatched bone marrow chimeric mice and sensitive DNA amplification techniques for the male-specific *SRY* gene have demonstrated that peripheral blood fibrocytes arise from either a radioresistant, bone marrow progenitor cell population or from other host tissue sources [8].

Given that fibrocytes rapidly enter sites of tissue injury, it is not surprising that these cells produce a variety of cytokines that serve to coordinate the successive inflammatory and reparative responses. DNA amplification analysis of mRNA obtained from fibrocytes in wound chambers has shown that these cells are an especially abundant source of growth factors, cytokines, and matrix components [14]. Fibrocytes express high levels of mRNA for the fibrogenic growth factors, platelet-derived growth factor-A (PDGF-A) and transforming growth factor-ß1 (TGF-ß1), the hematopoietic growth factor macrophage-colony-stimulating factor (M-CSF), and the chemokines macrophage inflammatory protein-1α (MIP-1α) and MIP-2. Fibrocytes also express low but detectable levels of mRNA for the pro-inflammatory cytokines, interleukin-1α (IL-1α) and tumor necrosis factor-α (TNF-α). Finally, among the various cell populations present in wound chambers, only fibrocytes express mRNA for Type I collagen.

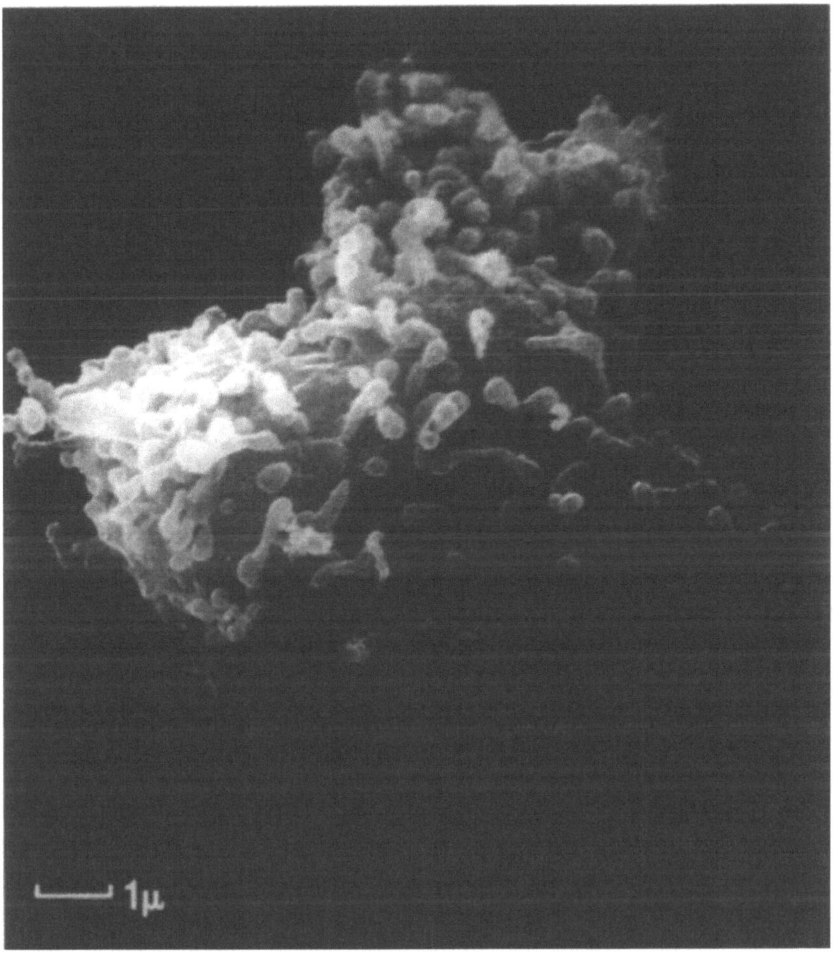

Figure 1. Scanning electron micrograph of a peripheral blood fibrocyte.

Recent studies suggest that fibrocyte collagen production is tightly regulated in the context of the inflammation and wound healing responses. For instance, the addition of the critical wound healing mediator IL-1ß to fibrocytes in culture suppresses Type I collagen production [14]. Conversely, IL-1ß induces fibrocyte secretion of the inflammatory chemokines MIP-1α, MIP-1β, monocyte chemoattractant protein-1 (MCP-1), IL-8, and GROα, the hematopoietic growth factors IL-6 and M-CSF, and the fibrogenic cytokines TGF-ß1 and TNF-α. IL-1ß may function to maintain peripheral blood fibrocytes in a pro-inflammatory state early in tissue repair, resulting in the increased production of molecules that recruit and expand the inflammatory cell population within the wound environment.

The skin is a vital barrier to infection or tissue invasion and plays a major role in host immunity [15]. When injured by physical trauma, burns, or vascular insufficiency, the skin can become a significant portal of entry for pathogenic microorganisms. Resident antigen-presenting cells (APCs), such as the Langerhans cell, initiate antigen-specific immune responses by processing and presenting microbial antigens to $CD4^+$ T cells via a class II major histocompatibility complex (MHC)-dependent pathway [16]. Recently, human peripheral blood fibrocytes were found to express each of the known surface components that are required for antigen presentation, including class II MHC molecules (HLA-DP, -DQ, and -DR), the co-stimulatory molecules CD80 and CD86, and the adhesion molecules CD11a, CD54, and CD58 [17]. Human fibrocytes also were found to be able to induce APC-dependent T cell proliferation when cultured with specific antigen. This proliferative activity was significantly higher than that induced by monocytes, and nearly as high as that induced by purified dendritic cells. Mouse fibrocytes also express the surface components required for antigen presentation (I-A, I-E, CD11a, CD54, and CD86), and function as potent APCs *in vitro*. In addition, mouse fibrocytes pulsed *in vitro* with antigen and delivered to a site of cutaneous injury were found to migrate to proximal lymph nodes and specifically to prime naive T cells. These studies suggest that fibrocytes may play an early and important role in the initiation of antigen-specific immunity.

The constitutive expression by fibrocytes of the surface proteins known to be necessary for antigen presentation contrasts with what has been described for tissue fibroblasts which require activation by interferon-gamma to express measurable quantities of HLA-DR [18]. Although several tissue-derived cells have been shown to be capable of presenting antigen to memory T cells, including dermal fibroblasts, endothelial cells, and melanocytes [18-20], the sensitization of naive T cells has been considered to be a particular function of dendritic cells [21-22]. Fibrocytes are distinct from dendritic cells and their precursors in their growth properties. for example, fibrocytes are an adherent, proliferating cell population whereas dendritic cells are non-adhering and poorly proliferating. They also differ in their surface protein expression ($collagen^+/CD13^+/CD34^+/CD25^-/CD10^-/CD38^-$). That fibrocytes also have a specialized and potent antigen presentation activity suggests that they may play a critical role in the initiation of immunity during tissue injury and repair.

The ability of fibrocytes to both recruit and activate T cells, and to secrete Type I collagen, suggests that these cells may play a critical role in certain connective tissue disorders. We hypothesize that a persistent fibrocyte:T cell activation response may lead to pathologically significant fibrosis in a variety of disease states. We recently examined fibrocyte function in Schistosomiasis, a parasitic disease characterized by a fibrosing, T cell-mediated reaction directed against parasite eggs that become entrapped in the hepatic and pulmonary circulations [23]. Immunohistochemical studies of livers obtained from *Schistosome japonicum*-infected mice have shown that numerous spindle-shaped $CD34^+$ cells co-localize to areas of connective tissue matrix deposition [14]. These data are the first to suggest that fibrocytes contribute to fibrotic pathology. Fibrocytes also may participate in the generation of excessive fibroses associated with various autoimmune disorders involving a persistent T cell activation, such as scleroderma or graft-versus-host disease.

2. TISSUE PROCUREMENT AND PROCESSING

Peripheral blood fibrocytes are readily harvested from the mononuclear cell fraction of whole blood. Procurement initially requires standard phlebotomy technique. Sterilize the skin overlying a target vein with 70% ethanol, place a tourniquet proximally, and puncture the target vein with a 18 to 21 gauge needle. Next, aspirate whole blood into a 60 ml syringe anti-coagulated with heparin sulfate and dilute 1:1 with sterile PBS. The mononuclear cell population is then fractionated by Ficoll-Paque gradient centrifugation. Under sterile conditions, place 30 ml of the diluted, heparinized blood into 50 ml conical centrifuge tubes, and slowly layer Ficoll-Paque solution (Pharmacia, Uppsala, Sweden) underneath the PBS/blood mixture by placing the tip of a pipet loaded with 12.5 ml Ficoll-Paque at the bottom of the conical tube. Disruption of the blood/Ficoll-Paque interface can be minimized by ejecting the solution slowly and evenly. Centrifuge the two phase samples for 30 minutes at 400*g*, 25°C with no brake. Using a sterile pipet, collect the mononuclear cell layer from the interface between the two phases and transfer it to another 50 ml conical centrifuge tube. Wash the mononuclear cells with excess PBS and centrifuge at 100 *g* for 10 minutes at 18-20°C. Aspirate the supernatant, resuspend the cells in PBS, and repeat the wash once to remove contaminating platelets. Resuspend the washed cell pellets in DMEM (Life Technologies, Gaithersburg, MD) supplemented with 20% fetal bovine serum (FBS) (Hyclone Labs, Logan, UT), and determine the concentration of the resultant cell suspension with a hemacytometer.

3. CULTURE TECHNIQUES

Peripheral blood fibrocytes differentiate from precursor mononuclear cells during long-term *in vitro* culture [8, 14, 17]. Load 2 ml of mononuclear cells suspended in DMEM supplemented with 20% FBS (2.5 x 10^6 cells/ml) into each well of a 6-well fibronectin- or collagen Type I–coated plate (Biocoat Human Cellware, Bedford, MA) and incubate at 37°C in 5% CO_2. After 24 hours of culture, gently aspirate the medium and replenish with pre-warmed DMEM supplemented with 20% FBS (Figure 2). Within 7-14 days of culture, a population of adherent, spindle-shaped cells will begin to differentiate and proliferate (Figure 3). After the fibrocytes reach about 75% confluency, gently aspirate the nonadherent cells and wash the adherent fibrocytes twice with 25°C PBS. Peripheral blood fibrocytes can be harvested by incubation in either ice-cold 0.05% EDTA in PBS, or commercially available trypsin-EDTA mixtures (Gibco).

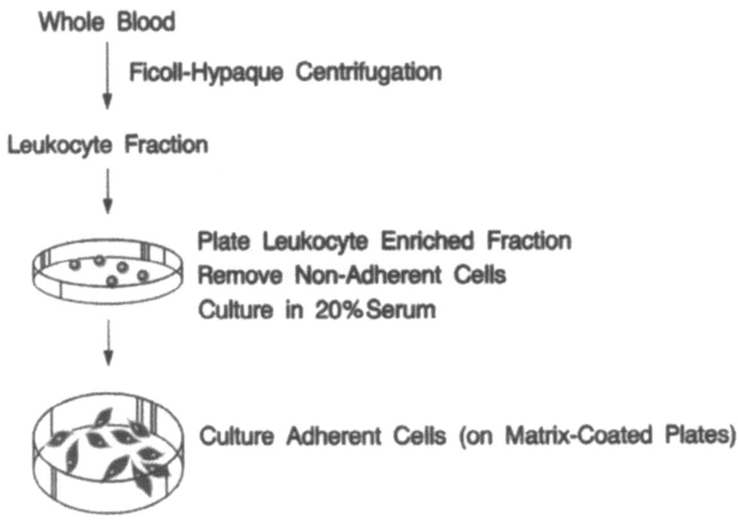

Figure 2. Isolation of peripheral blood fibrocytes.

Purity of the peripheral blood fibrocyte harvest can be increased from about 70-75% to more than 95% by immunomagnetic depletion of contaminating cell populations. This is accomplished with Dynabead magnetizable polystyrene beads (Dynal, Oslo, Norway) that are coated with primary monoclonal antibodies specific for T cells, monocytes, or B cells. Following the manufacturer's protocol, treat the fibrocyte preparation serially with conjugated beads that are specific for contaminating T cells (Dynabeads M-450 pan-T, αCD2), monocytes (Dynabeads M-450 αCD14), and then B cells (Dynabeads M-450 Pan-B, αCD19). For each depletion, resuspend the fibrocytes to 0.5 x 10^7 cells/ml in PBS/2% FBS, add $1x10^7$ beads/ml, and incubate for 30 minutes at 4°C on an apparatus that both tilts and rotates. After 30 minutes, add PBS/2% FBS to the cell-bead mixture to obtain a total volume of 10 ml, and then place in a Dynal magnet to remove the rosetted, contaminating cells. Wash the resultant purified fibrocytes twice in PBS and count with a hematocytometer.

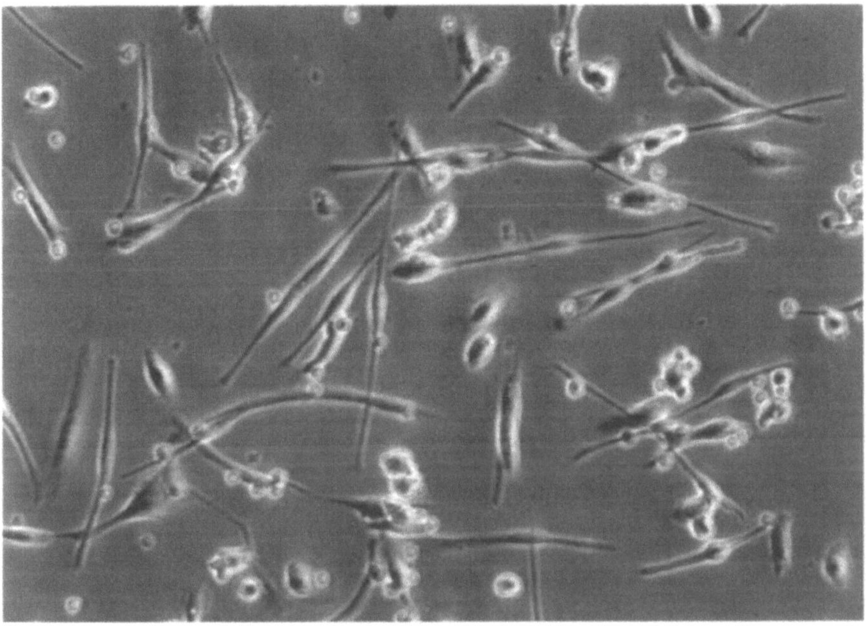

Figure 3. Photomicrograph of a peripheral blood fibrocyte culture after incubation for 10 days on a fibronectin-coated plate.

4. ASSAY TECHNIQUES

To verify that cultures of fibrocytes purified from human blood are homogeneous, the suspended cells are examined by flow cytometry for the coexpression of Type I collagen and CD34, a phenotype unique to fibrocytes [8]. Typical peripheral blood fibrocyte preparations are about 70% pure, and after negative immunoselection of T cells, monocytes, and B cells, the purity increases to more than 95% (Figure 4).

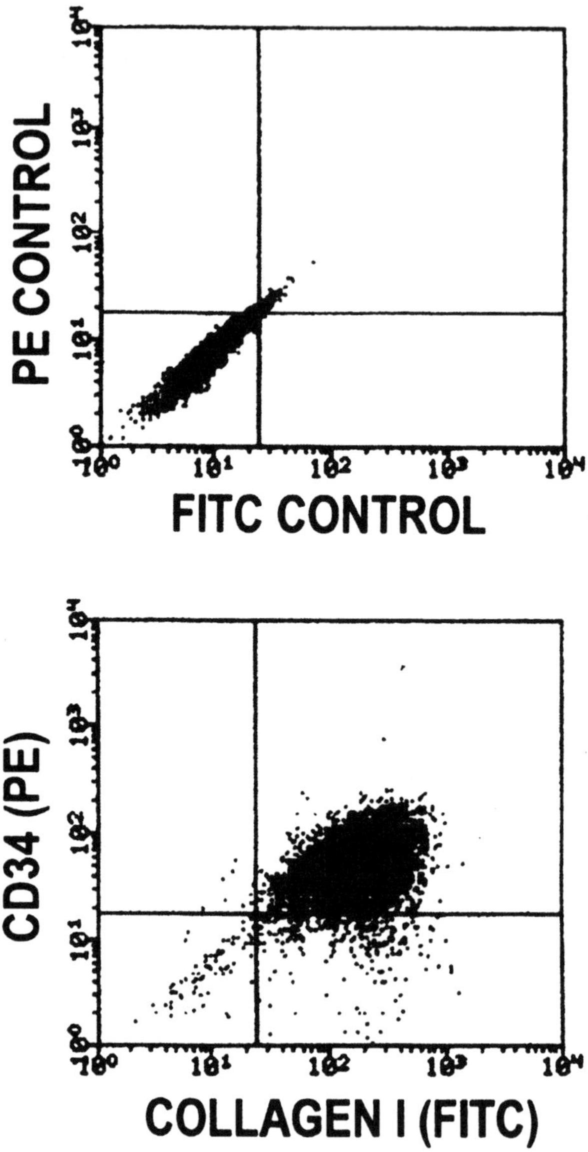

Figure 4. Type I collagen and CD34 co-expression by peripheral blood fibrocytes.

Wash 2 x 10^5 purified fibrocytes/sample twice in PBS containing 0.1% sodium azide (Sigma) and 1% bovine serum albumin (Sigma). Then resuspend the cells in 25 μl of diluted antibody (in PBS) and incubate for 30 minutes on ice. Wash the cells twice and resuspend in 200 μl of PBS. Analyze at least 10,000 cells/sample on a FACScan instrument (Becton Dickinson). Stain the cells with both a phycoerythrin-conjugated anti-CD34 mAb (clone 8G12) (Becton Dickinson) and a fluorescein-conjugated anti-collagen I mAb (clone MAB1340) (Chemicon, Temecula, CA). For controls, directly-conjugated isotype controls and cell only samples should be analyzed with each antibody.

5. UTILITY OF SYSTEM

Prior studies have demonstrated that fibrocytes rapidly enter sites of tissue injury and produce a variety of cytokines and connective tissue proteins [14]. The establishment of long-term fibrocyte cultures allows for the investigation of their functional characteristics *in vitro*. Cultured fibrocytes can be examined for their ability to: 1) secrete mediators (*e.g.* growth factors) and connective tissue proteins; 2) migrate on a variety of matrices; and 3) proliferate in response to mitogenic stimuli. Taken together, studies of fibrocyte function *in vitro* should further our understanding of the cellular and molecular mechanisms of wound healing.

Peripheral blood fibrocytes play an early, critical role in the initiation of antigen-specific immunity [17]. The capacity of cultured fibrocytes to activate $CD4^+$ and $CD8^+$ T lymphocytes, and their requisite cell surface and secretory interactions, may now be examined. The relative role of fibrocytes during the initiation of immunity *in vivo* has not yet been clarified, and thus *in vitro* functional comparisons with other antigen presenting cells can also be studied. Furthermore, examination of the homing characteristics of cultured fibrocytes introduced into the lymph system *in vivo* should provide insight into their relative role during T cell activation.

The establishment of *in vitro* cultures of peripheral blood fibrocytes makes feasible a wide range of clinical opportunities. Given their proliferative and migratory characteristics, these cells may be utilized for the targeted delivery of proteins via gene therapy. For example, cultured fibrocytes can be transfected with a potent growth factor, and then reintroduced into a person suffering from inadequate wound healing (*e.g.* diabetic patients). Based on their reparative homing characteristics [8], the transfected fibrocytes have the potential to target the secretion of the transfected growth factor to tissues that are undergoing critical repair processes. Cultured fibrocytes can also potentially be clinically employed to establish *in vivo* immunity against infectious agents (*e.g.* HIV) or neoplastic processes. For example, *in vitro* cultures of autologous fibrocytes can be pulsed with HIV-specific antigens (*e.g.* gp120, p24) and then re-introduced into HIV^+ patients in order to prime naive $CD4^+$ T lymphocytes, and thus induce anti-viral immunity. Alternatively, antigen-specific immunity may be inducible through the infusion of cultured fibrocytes that have been transfected with target antigens *in vitro*.

6. CONCLUDING REMARKS

The peripheral blood fibrocyte is a novel cell type that has been shown to rapidly enter sites of tissue injury and contribute to connective scar formation. Fibrocytes express a distinct profile of cytokines and growth factors and may function to attract and activate inflammatory and connective tissue cells. Fibrocytes also are specialized to present antigen, and may play a critical role in the initiation of cognate immunity during the earliest phases of tissue injury. In situations such as ischemia or diabetic vasculopathy, fibrocyte entry into damaged tissue sites may be compromised, thus contributing to poor scar formation. Conversely, peripheral blood fibrocytes may play a role in various pathological processes characterized by excessive fibrosis, including pulmonary and hepatic fibrosis, atherosclerosis, glomerulosclerosis, and pannus formation. Further investigation into the function of this novel cell population may provide important insight into the regulation of host wound healing and fibrotic responses.

REFERENCES

1 Morgan, C. J., and W. J. Pledger. (1992) in Wound Healing: Biochemical and Clinical Aspects. (Cohen, I. K., Diegelman, D. F., and Lindblad, W. J., editors), pp. 40-62, Saunders, Philadelphia.
2. Kovacs, E. J. and DiPietro, L. A. (1994) FASEB J. 8, 854-861.
3. Paget, J. (1863) Lectures On Surgical Pathology Delivered at the Royal College of Surgeons of England, 848 pp., Longmans, London.
4. Dunphy, J. E. (1963) N. Engl. J. Med. 268, 1367-1377.
5. Jackson, D. S. (1961) in The Biology of the Connective Tissue Cells, (Jackson, D. S., editor) pp. 172-178, Arthritis and Rheumatism Foundation, New York.
6. Petrakis, N. L., Davis, M., and Lucia, S. P. (1961) Blood 17, 109-118.
7. Labat, M. L., Bringuier, A. F., Arys-Philippart, C., Arys, A., and Wellens, F. (1994) Biomed. Pharmacother. 48, 103-111.
8. Bucala, R., Spiegel, L. A., Chesney, J. A., Hogan, M., and Cerami, A. (1994) Mol. Medicine 1, 71-81.
9. Diegelmann, R. F., Lindblad, W. J., and Cohen, I. K. (1986) J. Surg. Res. 40, 229-237.

10. Diegelman, R.F., Lindblad, W. J., Smith, T. C., Harris, T. M., and Cohen, I. K. (1987) J. Leuk. Biol. 42, 667-672.
11. Freudenthal, P.S., and Steinman, R. M. (1990) Proc. Natl. Acad. Sci. (U.S.A.) 87, 7698-7702.
12. Galy, A., Travis, M., Chen, D. and Chen, B. (1995) Immunity 3, 459-473.
13. Barclay, A.N., Beyers, A. B., Birkeland, M. L., Brown, M. H., Davis, S. J., Somoza, C. and Williams, A. F. (1993) The Leukocyte Antigen Facts Book, pp. 142-143, Academic Press, New York.
14. Chesney, J., Metz, C. M., Stavitsky, A. B., Bacher, M., and Bucala, R. (1998) J. Immunology 160, 419-425.
15. Mast, B. (1992) *In* Wound Healing: Biochemical and Clinical Aspects, (Cohen, I. K., Diegelman, R. F., and Lindblad, W. J., editors), pp. 344-355, Saunders, Philadelphia.
16. Chu, T. and Jaffe, M. B. (1994) Br. J. Cancer Suppl. 23, S4-S10.
17. Chesney, J., Bacher, M., Bender, A., and Bucala, R. (1997) Proc Natl Acad Sci 94, 6307-6312.
18. Geppert, T. D. and Lipsky, P. E. (1985) J. Immunol. 135, 3750-3762.
19. Pober, J.S., Gimbrone, M. A., Cotran, R. S.,Reiss, C. S., Burakoff, S. J., Fiers, W., and Kenneth, A. (1983) J. Exp. Med. 157, 1339-1353.
20. Poole, I.C., Mutis, T., Rene, M., Wijngaard, G. J., Westerohof, W., Ottenhoff, T., Vries, R., and Das, P. K. (1993) J. Immunol. 12, 7284-7292.
21. Inaba, K., Metlay, J. P., Crowley, M. T., and Steinman, R. M. (1990) J. Exp. Med. 172, 631-640.
22. Levin, D., Constant, S., Pasqualini, T., Flavell, R., and Bottomly, K. (1993) J. Immunol 12, 6742-6750.
23. Warren, K. S., Domingo, E. O., and Cowan, R. B. T. (1967) Am. J. Pathol. 51, 735-756.

Chapter 11

Osteoblasts

Lucy Di-Silvio and Neelam Gurav
Institute of Orthopaedics (Royal Free and University College Medical School), Royal National Orthopaedic Hospital Trust, Stanmore, Middlesex, HA7 4LP, UK. Tel: 0044-208-909-5825, Fax: 0044-208-954-8560; E-mail: l.disilvio@ucl.ac.uk

1. INTRODUCTION

This chapter describes a primary human osteoblast (HOB) cell model, focusing on culture conditions that favor proliferation and differentiation. Biochemical, light and electron microscopic methods of characterization of growth and matrix production are also given.

The method described shows that it is possible to select for osteoblastic cells and achieve high yields of phenotypically stable HOBs. This finding is in contrast to the suggestion that osteoblastic characteristics are lost rapidly in culture [1].

1.1 Rationale for the use of isolated bone cells

Numerous *in vitro* models using human and animal primary, immortalized and transformed osteoblastic cells have been described. Primary cultures of osteoblast-like cells are advantageous for studies of bone cell metabolism and differentiation because they retain a normal genotype [2-5].

A hypothesis for differentiation of the marrow stromal compartment analogous to that in the hematopoietic lineage was proposed [6], where stromal cell lines give rise to committed progenitors for different cell types [7-8]. Osteogenic cell subpopulations can be derived from colonies of bone marrow stromal cells that grow as fibroblastic colonics initially, but form calcified tissue resembling bone when implanted in diffusion chambers *in vitro* [9]. Bone formation under specific culture conditions in diffusion chambers with cultured human bone-derived cells and cultured marrow stromal cells has also been reported [10].

M.R. Koller, B.O. Palsson and J.R.W. Masters (eds.), Human Cell Culture Vol V, 221–241.

Cells grow from explants of normal adult bone in culture [11-13]. The ability of osteoblasts to migrate from bone explants from various species has been reported, and some of the methods have been adapted for use with human bone [12, 14]. The cells appear to be mainly osteoblastic, although it has proved difficult to isolate a homogenous primary cell population that is able to retain the characteristic phenotype *in vitro* [15]. Isolation techniques for osteoblasts have been compared and the cells vary in the amounts of osteocalcin and type 1 collagen they produce [16]. The region the cells are isolated from, as well as the method of isolation, play a significant role in the behavior of the osteoblast [17].

The ideal system for HOBs should allow long-term culture of cells without de-differentiation. Under selected conditions, it is possible to grow differentiated cells that exhibit tissue-specific metabolic responses and synthesise structural matrix components [18]. The *Osf2/Cbfa 1* genes have major regulatory roles in osteoblast differentiation [19-21]. Absence of the Cbfa1 gene leads to a lack of ossification. This transcription factor has an important role in osteoblast differentiation and has been shown to induce expression of osteocalcin and osteopontin [22].

1.1.1 Comparison of human and animal cell lines

Continuous cell lines derived from rat, mouse or human osteosarcoma, such as UMR 106, ROS 17-2.8, SOS, U-2OS, MC3 T3-E1, MG-63 and HOS TE-85, are readily available and widely used [23]. Differences have been noted within species and between transformed and non-transformed cells [24, 25]. These cells are not analogous to adult human bone cells and can express different receptors *in vitro*. They often have different proliferation rates and express varying levels of ALP activity and osteocalcin in response to $1,25(OH)_2D_3$ [26].

1.1.2 Comparison of primary cells and cell lines

Several methods have been described for transforming human osteoblasts [2, 26-28]. However, the cells have abnormal growth characteristics and the degree of maturation and responsiveness of the fetal cells is not comparable to those of adult human bone cells.

Clonal cell lines may express receptors which are not normally expressed [24,26]. This is of particular relevance in studies investigating the effect of growth factors, since transformed cells may respond differently to normal cells and may produce different growth regulators. In addition to heterogeneity between clones, phenotypic variations can exist between daughter cells derived from a single clone [18]. These differences may occur as a result of the cells being exposed to different growth conditions at different stages in the cell cycle. A high or low cell density or the presence or absence of mineralizing agents such as ß-glycerophosphate can also cause differences. At a high density, for example, it is believed that osteoblastic expression is increased, whereas at low density these cell markers are reduced [23, 29].

In general, immortalized HOBs exhibit osteoblastic properties and characteristics, including high ALP, response to PTH and $1,25(OH)_2D_3$, production of osteocalcin and collagen type I and, in some cases, *in vitro* mineralization [5, 30, 31]. The effects of $1,25(OH)_2D_3$ on alkaline phosphatase activity, cell growth and cell protein in 'pre-osteoblastic' and 'mature osteoblastic' cultures varied depending on the stage of osteoblastic maturation [32].

The progressive de-differentiation is a major problem with clonal lines [33]. The production and accumulation of osteoblast products in the extracellular matrix and changes in cell shape provide signals for differentiation. The acquisition of normal osteogenic cell morphology may be a prerequisite for further development and maturation [13, 34].

1.1.3 Mineralization

Bone development can be divided into three stages: proliferation, extracellular maturation and mineralization [35]. Each of these stages is reflected by peaks in gene expression levels of specific proteins. In order for bone cells to pass from one stage to the next, specific signals are needed. Mineralization can be divided into two phases. In the first phase, osteoblasts secrete an organic matrix which is considered to be a preosseous matrix, and is usually termed osteoid. The osteoid consists of type I collagen and other proteins including proteoglycans, glycoproteins and non-collagenous proteins. During the second phase, mineralization occurs, transforming the osteoid into bone [36]. In the HOB culture system described in this chapter, three distinct phases occur: growth (proliferation), matrix development and mineralization, recapitulating the stages seen *in vivo*.

The process of mineralization in lamellar bone and the factors controlling it are poorly understood, one reason being the difficulty in obtaining and maintaining enough primary cells. Another reason is the frequent use of fetal bone, which has growth characteristics that are different from those of adult bone [2, 27, 37]. Mineralization studies have used osteoblastic cells derived from human and rodent bone tissue, and in the majority of cases these have been derived from embryonic or neonatal bone or marrow [2, 11-14, 23, 30].

Osteoblastic cells can be grown as compact cell colonies and then calcification initiated by a variety of means, including suspension of cells in agarose, methylcellulose and collagen type I [38-40]. It has been suggested that a prerequisite for osteoblast mineralization is the formation of multilayers. This can be achieved by seeding cells in a micromass [41], where a large number of cells are concentrated into a small volume of medium and seeded into a dish or a plate. The high concentration of cells in such a small area causes the cells to adhere closely to one another, sometimes on top of each other forming layers. However, spontaneous differentiation and calcification can occur with the deposition of hydroxyapatite in well-developed bone matrix in monolayers of mouse MC3T3-E1 cells [4]. Osteoblasts have also been cultured in native collagen type I gel and been shown to mineralize [42].

Numerous reports have described bone cell culture methods, but few have shown spontaneous *in vitro* mineralization [5, 43]. In all reported cases, certain factors or conditions have been shown to be a necessary prerequisite for terminal cell differentiation and mineralization. Examples include (1) three-dimensional cell culture [41], (2) Na-ß-glycerophosphate or other organic phosphates as stimulators [38, 43, 44] and (3) enrichment of the nutrient medium [5, 29, 41].

In primary cultures, isolated osteoblasts have been shown to synthesize several proteins and enzymes which are known to be localised in bone, including alkaline phosphatase (ALP), osteocalcin and type I collagen. Primary human bone cells in conventional culture reach confluence in 3-6 weeks. If maintained in culture for sufficient lengths of time, the cells form 'nodules', and in these areas the cells lay down a dense matrix and develop a granular appearance. These nodules appear to consist of calcium phosphate crystals embedded in matrix. The process is usually associated with the release of vesicle-type structures from the cell surface and this may initiate mineral deposition. In some culture systems the development of crystals can be accelerated by the addition of 5-10 mM β-glycerophosphate. In the culture system described in this chapter, nodule formation occurs spontaneously after approximately 30 days in culture with a dense granular matrix which stains positive for calcium.

Culture conditions contribute significantly towards the ability of cells to differentiate and calcify. The addition of agents such as calcium ß-glycerophosphate, glucocorticoids, sodium ß-glycerophosphate, calcium hexose monophosphate and dexamethasone has been used to promote mineralization of osteoblasts in culture [14, 41, 43, 45, 46].

2. TISSUE PROCUREMENT AND PROCESSING

Clinical applications such as bone healing and reconstruction require a supply of bone cells. Several methods of obtaining bone cells have been documented, all yielding varying results depending on the clinical history. The three most common methods are bone marrow stromal cell cultures and isolation from bone explants, with or without collagenase digestion. The first uses growth factors to induce cells to follow an osteoblast pathway and the second involves the migration of cells from tissue explants.

The method in this chapter describes the isolation of osteoblasts from bone explants with collagenase digestion. Femoral heads were obtained from patients undergoing surgery for total joint replacement. The success rate of the extraction procedure is over 80%, with loss occurring mainly due to infection. Other factors determining success rate include the age of patient, clinical status and time elapsed from resection of tissue to processing and culture.

As bone is a hard tissue, it is advisable to wear a chain-mail glove underneath the sterile glove on the hand holding the sample. The bone specimen can be processed up to several hours following excision. However, the sooner the tissue is processed the better the yield of cells. If absolutely necessary, the tissue can be stored at 4°C overnight in unsupplemented Dulbecco's Modified Eagles Medium (DMEM) (Sigma, Poole, England) at 4°C.

3. CULTURE TECHNIQUES

The bone sample is placed in a 90 mm tissue culture dish and trabecular bone fragments dissected under sterile conditions. Trabecular bone is cut, scraped or teased out of the femoral head with a scalpel blade or forceps. The bone fragments are placed into a sterile universal container and washed thoroughly in calcium and magnesium-free phosphate buffered saline solution (PBS). By capping the universal and shaking it vigorously, red blood cells, marrow and tissue debris from the bone fragments are removed. The washing procedure is repeated several times, each time changing the PBS until the liquid surrounding the fragments has cleared as much as possible.

The washing steps are important for the removal of blood, fat and debris and the surface layer of cells, and also to expose the trabecular surfaces of the bone fragments. The bone fragments are given one final wash in osteoblast medium and transferred to a clean petri dish where the fragments are chopped as finely as possible with sterile scalpel blades. The osteoblast medium consists of DMEM containing 10% fetal bovine serum (FBS), 1% non-essential amino acids, 150 µg/ml ascorbic acid, 2 mM L-glutamine, 0.01 M (4-(2-hydroxyethyl)-1-piperazineethanesulfonic acid) (HEPES), 100 units/ml penicillin and 100 µg/ml streptomycin.

The fragments are divided between two or three 60 mm dishes, depending on the quantity of tissue. Five ml of osteoblast medium is added and the dishes placed in an incubator at 37°C with 5% carbon dioxide in a humidified atmosphere. The bone fragments should be inspected daily to check for release of blood and marrow cells (rounded, non-adherent). If these are present, the medium is changed and replaced with fresh medium. This 'pre-digestion' explant incubation period is important as it allows the migration and removal of marrow cells, blood cells and other non-adherent cells. During this period cells, migrate from the bone fragments forming an osteoid seam around the explant. When osteoid seam formation is good and there are many cells adhering to the tissue culture dish, osteoid digestion is carried out in order to harvest the cells. Cell migration is usually observed in over 80% of the bone fragments after 4-5 days, with seams formed around the fragments.

The bone fragments are removed from the petri dish and placed into a sterile universal container containing enzyme mixture, consisting of collagenase (100 U/ml), and trypsin (300 U/ml) (Sigma, England) in PBS buffered with 0.01 M HEPES. This is placed on a rotating mixer at 37°C for 20 minutes, after which the fluid in the universal becomes cloudy, indicating that the cells have been released into the solution. Fragments are incubated for 20 minutes. A shorter time period results in a more mixed cell population and a longer incubation results in cell damage and loss of viability.

Following the 20 minute incubation, the supernatant is carefully transferred to a fresh universal and centrifuged at 2000 rpm at 18°C for 5 minutes. There should be a cell pellet at the bottom of the universal at this stage. The supernatant is removed with a sterile pipette and discarded. Two ml of osteoblast culture medium is added to the pellet, which is resuspended using a pasteur pipette. The volume is then made up to 10 ml with osteoblast medium and centrifuged as before. Using a small volume of medium to resuspend the cells results in a uniform cell suspension and can obviate the need to use a needle and syringe to obtain a single cell suspension. The supernatant is again removed with a sterile pasteur pipette, taking care not to disturb the cell pellet, and the pellet is re-suspended in 2 ml of osteoblast medium as before. If the cell pellet does not disperse fully to form a single cell suspension, a 5 ml syringe with a 23 G hypodermic needle can be used to disaggregate the cells. This requires gently "forcing" the cell suspension through the needle several times to produce a single cell suspension. Care should be taken not to cause foaming of the medium and consequent damage to the cells.

Using this method, the cell yield is usually between 10^5-10^6 cells. These are plated in one 25 cm^2 flask with a total of 5 ml of osteoblast medium. If the yield of cells is greater (10^6-10^7), the cells can be seeded into more flasks. The seeding density should be approximately one million cells per 25 cm^2 flask. If the surface area of the flask is too large for the number of cells isolated, a smaller flask should be used. These cells constitute passage one.

The morphology of the primary cells five days following enzyme digestion is typically osteoblastic. Initially, the cells have a polygonal morphology that becomes cuboidal when the cells reach confluence, which can occur between days 5-14 depending on the initial seeding density (Figure 1a-f). Under normal conditions, the HOBs display logarithmic growth with a doubling time of approximately 36 hours. Once sufficient cells have been cultured, they are characterized by measuring cell viability, doubling time and phenotypic markers. In our experience, these cells have maintained phenotypic stability following cryopreservation up to passage 19. The preparation procedure and the inclusion of a pre-digestion incubation period reduce contamination with marrow cells and allow the selection and proliferation of a predominantly osteoblastic cell population.

When cells from passage one have reached confluence, it is necessary to trypsinise and split the cells. It is advisable to grow sufficient cells from each passage to ensure that at least one vial can be cryopreserved. For trypsinization, the medium is removed and 2-5 ml of PBS warmed to 37°C is added, gently swirled around and then removed.

This wash procedure removes serum components which would otherwise block thc action of the enzyme. Following the PBS wash, 2-5 ml of trypsinization solution (300 U/ml trypsin with 0.01 M HEPES in PBS) is added and the cells incubated at 37°C for approximately 5 minutes. The progress is monitored at each stage under the microscope and the cells should appear rounded and start detaching from the flask. At this stage, a gentle tap of the flask should dislodge most of the cells. If cells remain adhered to the surface, the flask should be placed back in the incubator until all the cells have dislodged.

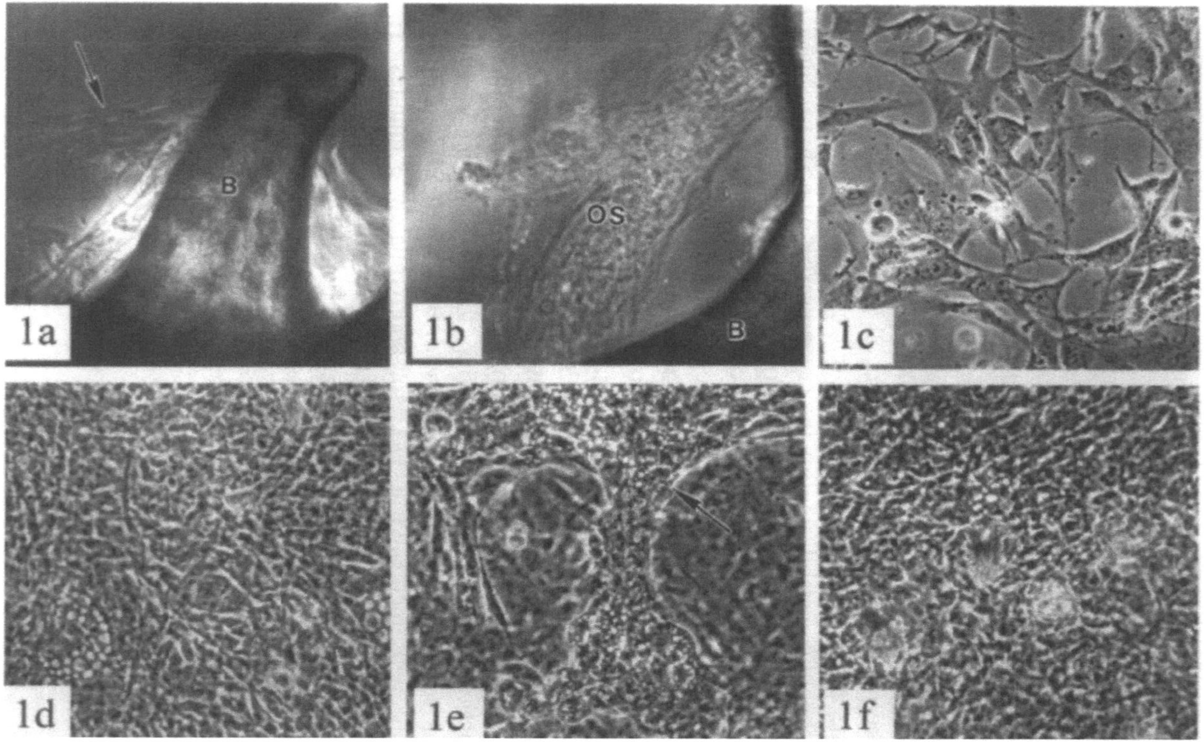

Figure 1a. HOB cells, (arrow) migrating from a bone fragment (B) after 3 days in culture. Fig 1b: Osteoid Seam (OS) seen forming around a bone fragment after 5 days in culture. Fig 1c: Appearance of HOB cells in culture 5 days post enzymatic digestion of bone fragments. Fig 1d: HOB cells after 12 days in culture showing confluent layer of cells. Fig 1e: HOB cells after 21 days in culture; note the densely packed cells forming multilayers (arrow). Fig 1f: HOB cells after 24 days in culture; the cells have formed layers with the appearance of bone nodules, which become mineralized by day 28 in culture.

When all the cells have detached, 5 ml of osteoblast medium is added and collected into a sterile universal followed by another wash with 5 ml of medium. The cells are centrifuged at 2000 rpm for 5 minutes and the medium removed using a sterile pasteur pipette, taking care not to disturb the cell pellet. The cell pellet is resuspended in 10 ml of medium and centrifuged again as above. The medium is removed and discarded as before, and the pellet resuspended in 2 ml fresh medium. If required, a 5 ml syringe with 23 G needle can be used to obtain a single cell suspension. The volume is made up to 10 ml and divided between the required number of flasks with extra medium added according to the volume of the flask.

Generally speaking the cell yield from one confluent 25 cm^2 flask is between 1-2 x 10^6 cells. Therefore, for sub-culturing purposes the cells from one confluent 25 cm^2 flask can be placed in two-three 25 cm^2 flasks. These new flasks will take 3-7 days to reach confluence, depending on the quality of the cells. At this stage the cells can be seeded into larger flasks. It is vital to check the cells every day. Medium should be changed at least twice a week for long-term, slow-growing cultures.

For cryopreservation the cells are frozen in a mixture of FBS and dimethylsulfoxide (DMSO). Following trypsinisation and centrifugation as described above the medium is removed and discarded. The pellet is resuspended in 1 ml of freezing mix (10% DMSO in FBS), using a 2 ml syringe with 23 G needle, if required, to produce a single cell suspension. The vials are left at –70°C overnight for the first stage of the freezing process and then transferred into liquid nitrogen. It is important to ensure that the vials are transferred to liquid nitrogen within 24 hours of initial freezing, as leaving them at –70°C for longer periods results in a loss of viability.

In order to resuscitate cryopreserved cells, osteoblast medium is warmed to 37°C and the vial of cells to be resuscitated removed from liquid nitrogen and placed in a laminar flow cabinet. Once the liquid begins to thaw it is transferred immediately into 10 ml of osteoblast medium. The cell suspension is centrifuged at 2000 rpm for 5 minutes and the liquid is removed, taking care not to disturb the cell pellet. Another 10 ml of medium is added to the cells, which are then mixed and recentrifuged. The medium is then removed and discarded and the pellet resuspended with 2 ml fresh medium using a 5 ml syringe with 23 G needle, in order to produce a single cell suspension. The volume is made up to 10 ml and the cell suspension divided between the required number of flasks, culture dishes or plates.

4. ASSAY TECHNIQUES

4.1 HOB Characterization

In order to characterize the HOB cells, various analyses need to be performed. These include measurement of cell growth, alkaline phosphatase (histochemical staining and biochemical measurement), osteocalcin and procollagen type 1 (PICP), and response to PTH by measuring cyclic adenosine monophosphate (cAMP) production. These methods are described below.

4.1.1 Alkaline Phosphatase

Alkaline phosphatase is used as a marker of osteoblast phenotype. Alkaline phosphatase has been implicated in the initiation of mineralization, but histochemically in cultured cells, only a fraction of the cells stain positively, even in clonal lines. Using immediate digestion, monolayer cultures of normal human bone cells contained subpopulations of alkaline phosphatase positive cells [47] and, from our experience, this approach results in a mixed cell population with predominantly immature precursors of osteoblasts.

Only a proportion of cells undergo maturation in culture, and this proportion may differ amongst different clonal lines. ALP activity in human bone cell cultures is dependent on cell cycle distribution, cell density and length of time in culture [48]. The timing of the phases that accompany differentiation *in vitro* differ depending on the method used, resulting in a lag between the different phases of expression of the individual markers [49]. Therefore, the differences reported in the literature may be a result of the different techniques used to isolate bone cells and could explain the variations seen in terms of cell confluence, differentiation, matrix formation and mineralization.

Alkaline phosphatase activity in cell lysate and medium can be determined using a COBAS-BIO (Roche, UK) centrifugal analyser. The assay measures the release of p-nitrophenol from p-nitrophenol phosphate (PNPP) at 37°C in buffer containing 1 M diethanolamine (DEA), 10 mM PNPP and 0.5 mM $MgCl_2$ and 0.22 M NaCl, pH 9.8 (Merck, UK). The production of ALP by HOBs is detectable from approximately day 4 onwards, rising to a peak from day 10 onwards. A rapid fall in cell proliferation is coincident with a rapid increase in ALP production.

Histochemical detection of alkaline phosphatase can also be performed using a modification of a published method [50]. The incubation mixture consists of 4% New Fuchsin in 2 M HCl plus an equal volume of 4% sodium nitrite, which are shaken together and then added to 40 ml 20 mM Tris, pH

9.0. To this is added 10 mg of the substrate napthol AS-BI-phosphate, sodium salt (Sigma, England). The solution is mixed and added to the dishes or dropped onto slides and incubated for 30 minutes at 37 °C. The medium is discarded and the cell layer washed with PBS, and then counterstained with methyl green for 2 minutes followed by a further wash in water. Finally, the specimen is drained and mounted in Aquamount (Merck, UK).

ALP activity can also be investigated using electron microscopy (Figures 2a and 2b). Cells are fixed in their culture dishes with 1.5% glutaraldehyde in 0.1 M sodium cacodylate buffer (pH 7.4) at 4°C for 1 hour, followed by a further 30 minutes in 0.1 M sodium cacodylate buffer only. The incubation medium is made up from the stock solutions: 0.045 M sodium ß-glycerophosphate, 0.025 M magnesium chloride and 0.018 M lead nitrate. These are aliquoted in 2 ml volumes and stored separately at -20 °C. Tris (hydroxymethyl) methylamine/HCl (Merck, UK) buffer at pH 9 is prepared fresh. A working solution of 10 ml is made up from the stock added in order as follows: 2 ml distilled water, 2ml 0.2 M Tris/HCl (pH 9.0), 2ml of the aliquoted 0.045 M sodium ß-glycerophosphate, 2 ml 0.025 M magnesium chloride and 2 ml 0.018 M lead nitrate. To test for false non-enzymatic deposition of reaction product, a control sample is incubated in a 2 mM levamisole hydrochloride (Aldrich, UK) solution to inhibit alkaline phosphatase activity. The incubation medium of 40 mM Tris/HCl, 9 mM sodium ß-glycerophosphate, 5 mM magnesium chloride and 3.6 mM lead nitrate is added to the culture dishes and incubated for 30 minutes at 37°C. The dishes are washed briefly in 0.1 M sodium cacodylate buffer (pH 7.4), post-fixed for 30 minutes in 1% osmium tetroxide in 0.1 M sodium cacodylate buffer and washed in several changes of buffer. The cells in the culture dish are dehydrated through a graded series of ethyl alcohol (70%, 90%, 96% and 100%, dried with sodium sulfate anhydrous).

The cells are removed from the surface of the tissue culture plastic by flooding the dish with propylene oxide. The dish must be continuously agitated until the entire cell layer is completely detached, which can be seen by naked eye. After detachment, the cell sheet is immediately lifted out of the dish using either fine forceps or a fine dissecting seeker. It is then placed into fresh propylene oxide for 5 minutes to remove any excess plastic. The cell sheet is infiltrated with Spurrs' resin for one hour, followed by a further two changes for one hour each in pure resin. The cell sheet is placed onto a slab of dental wax and chopped finely, a drop of resin is placed on top and then collected with a pasteur pipette into an eppendorf tube with fresh resin. The eppendorfs are centrifuged at 13,000 rpm for 5 minutes to obtain a firm pellet and the resin cured for 18 hours at 70°C and cut.

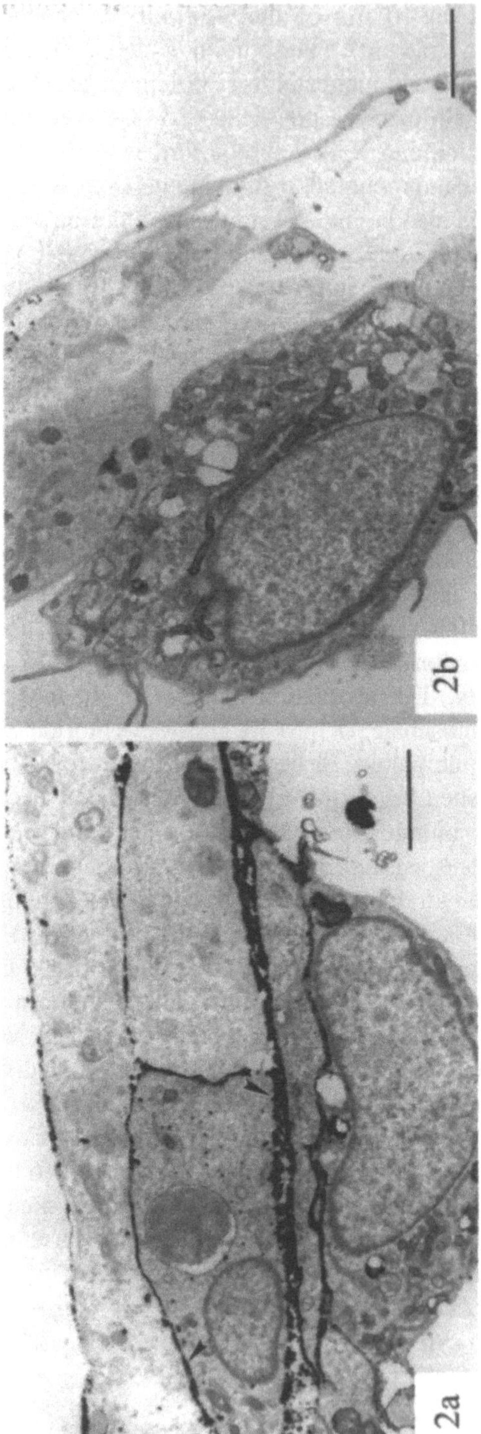

Figure 2a. Electron micrograph of a 12-day monolayer culture of HOB cells showing electron dense positive localisation of ALP along the cell membrane (arrowheads). Figure 2b: A negative control of a 12-day monolayer HOB culture where ALP activity was blocked by levamisole hydrochloride. (Bar = 2 µm).

4.1.2 Collagen

Collagen is the most abundant of the bone matrix proteins. Type I collagen derived from the larger protein, Type I procollagen, is an indicator of matrix synthesis. The carboxy terminal propeptide of type I procollagen (PICP) is used as a biochemical indicator of type I collagen production. PICP is measured using a radioimmunoassay (Orion Diagnostica, Pharmacia UK). Cross-reactivity with the carboxy terminal propeptide of type III procollagen has been minimized in this kit by purification of the antigen and selection of antiserum. Assay conditions are modified by increasing the first antibody incubation period to 24 hours at 4°C, in order to increase assay sensitivity. A functional relationship appears to exist between the down-regulation of proliferation and the initiation of extracellular matrix maturation as indicated by a sharp rise on day 8 in the production of PICP by HOB cells.

4.1.3 Osteocalcin

Osteocalcin, a vitamin K-dependent calcium-binding protein is synthesized only by osteoblasts, and is a sensitive marker of osteoblast differentiation and mineralization. Osteocalcin is measured by competitive radioimmunoassay (OSCA test, Henning, Berlin) using an antibody coated-tube technique and ^{125}I-labeled osteocalcin. Osteocalcin release into the medium is determined using basal medium as a control. Conventional assays are mainly for use with serum or plasma, hence they should be validated for use with tissue culture medium and cell lysate. The antibody used was raised against intact human osteocalcin and the assay was validated for tissue culture medium. The production of osteocalcin seems to be dependent on the extent of mineralization. However, low levels can be detected from as early as day 4 in our HOB preparation in monolayer culture. Immediately following down-regulation of proliferation, as reflected by a decline in DNA synthesis, the expression of ALP increases with osteocalcin expression following shortly thereafter [51].

4.1.4 Parathyroid Hormone (PTH)

Responsiveness to PTH is specific to osteoblastic cells. The HOB cells respond to PTH by an increase in intracellular cAMP production. The amplitude of the response to PTH can be variable, depending on the age of the cultures. Cyclic AMP (cAMP) production is measured in HOB treated with PTH (10^{-8} M) for 24 hours on days 3, 7 and 14 of culture. The PTH (bovine 77/533) is obtained from the National Institute for Biological

Standards and Control (South Mimms, UK). Intracellular cAMP is measured using the BIOTRAK kit (Amersham International, Amersham, UK) and adopting the non-acetylation method, suitable for culture medium. Under basal conditions, cAMP production is undetectable before day 3 for the HOB cell cultures, but usually increases after day 7.

4.1.5 Calcium deposition

HOB cells in monolayer spontaneously form nodules after approximately 21 days with calcium deposition in the absence of ß-glycerophosphate. To stain for calcium using Alizarin red S, cells can be fixed in neutral formalin, alcohol or formal alcohol. A 2% aqueous alizarin red S solution is prepared and the pH adjusted to 4.2 with 10% ammonium hydroxide. The sections are taken to water and then rinsed in distilled water before transfer to Alizarin red S solution for 1-5 minutes, blotting and rinsing in acetone for 30 seconds. The sample is then treated with acetone-xylene 1:1 for 15 seconds and rinsed in fresh xylene and mounted in DPX. Calcium deposits are stained orange-red and staining time depends upon the amount of calcium in the sections. Von Kossa's staining can also be used for detecting mineralization. Sections or cells on coverslips are transferred to distilled water and placed in 1% aqueous silver nitrate solution and exposed to strong light for 10-60 seconds. The section is then washed in 3 changes of distilled water and placed in 2.5% sodium thiosulphate for 5 minutes, thoroughly washed in distilled water, dehydrated and mounted in DPX. The mineralized portion appears black and the osteoid is stained red.

5. UTILITY OF SYSTEM

In normal bone, growth factor (GF) production and availability play key roles in bone formation and remodeling, and it is likely that locally produced GFs are intimately involved in the balance between resorption and accretion [52, 53]. The responsiveness of osteoblast-like cells to GFs *in vitro* has been studied mostly in rodent fetal cell cultures [54-56]. One of the most important GFs in osseous tissue development is growth hormone (GH). For example, the *in vitro* effect of GH on chondrogenesis and osteogenesis in mouse cartilage progenitor cells in promoting *de novo* bone formation has been demonstrated [57]. Recent studies have shown that GH can also affect both the proliferation [58] and differentiation of normal human osteoblasts [59, 60]. GH has been shown to increase biochemical markers of osteoblast activity, both as a direct effect on proliferation of osteoblasts and also an increase in the expression of proteins associated with differentiation [61].

Administration of GH to human volunteers increases the biochemical markers of bone formation, suggesting increased activity of osteoblasts and induction of the proliferation of their precursors [62]. GH has been shown to have a mitogenic effect on osteoblast-like cells [54, 63, 64], mediated by the paracrine or autocrine action of insulin-like growth factor-1 (IGF-1) [55, 60]. IGF-1 was able to support the differentiation of cultured osteoblast-like cells and increase alkaline phosphatase activity [65].

The HOB cell system described in this chapter provides an excellent model to study the effects of GFs on osteoblast cell proliferation and differentiation. Dose-dependent stimulation of primary HOB cells following exposure of the cells to GH and IGF-1 has been studied [61]. An optimal dose range to use for most GFs is between 50-150 ng/ml, as doses lower than this usually do not have an effect and higher doses result in down-regulation of receptors. However, the origin of the cells plays a significant role in their response to various stimuli. Osteoblasts removed from women were more susceptible to age changes as compared to cells isolated from men, and cells derived from post-menopausal women had a decreased number of osteoblasts, and in some cases were less able to differentiate [66].

The use of biomaterials to restore the function of traumatized or diseased joints has greatly increased the demand for the development of innovative bone analogue materials. Biomaterials are now being developed to elicit specific responses. The HOB model described in this chapter has been used extensively to study tissue-materials interaction [67-69].

Prior to implantation, all materials have to be tested for biocompatibility, where "the ability to perform with an appropriate host response in a specific application" is determined [70]. This allows the assessment of cytotoxicity. In addition, biofunctionality can be studied, to determine the response for a given application using the "appropriate" cell system [71-73].

The trend towards tissue-engineered procedures has focused on the development of appropriate polymer scaffolds, which can act as carriers for osteoinductive factors and support for cell-seeded systems. Bioactive composites are currently being investigated in our laboratory as potential bone substitute materials. An example includes HAPEX™, a hydroxyapatite reinforced polyethylene composite. This material *in vivo* has shown cancellous and cortical bone growth adjacent to the implanted ceramic surface in rabbit femoral condyles [74, 75].

Our *in vitro* studies indicate that HAPEX™ is an ideal template and provides a favorable site for the recruitment of cells from the surrounding biological environment [76-78]. Material surface topography plays a critical role in cell-biomaterial interactions, and we have shown that presentation of a bioactive surface results in a favorable attachment of cells and subsequent proliferation and differentiation [79].

Porous hydroxyapatite is being developed for tissue guidance and as a scaffold for tissue engineering. Ongoing studies in our laboratory have shown that both porosity and interconnectivity are important for the promotion of rapid bone ingrowth [80].

6. CONCLUDING REMARKS

Many methods have been described for the isolation of both animal and human primary osteoblasts. Primary cultures are favored for studying bone cell function and for investigating the effect of bone growth factors. The method we describe for the isolation of primary HOB results in a homogenous cell population with a stable phenotype.

Using other methods for culturing HOB, de-differentiation can occur. It is not known whether differences between clones are due to the presence of cells at different stages of maturation or arise as a result of de-differentiation in culture [23, 26, 81]. In a rat cell line, it was shown that cells can alter their phenotype and generate a number of sub-populations, each of which may generate further heterogeneity [82]. Variations in phenotype between progeny of cells within a single clone can occur [20, 29]. These changes are not limited to transformed cells, but have also been reported in normal osteoblasts [15, 17]. The use of characterized cell systems such as the one described in this chapter will provide fundamental basic information for advances in the repair of skeletal tissues.

ACKNOWLEDGMENTS

The authors wish to thank Mr. MV Kayser for his excellent histology and microscopy expertise, Mrs. Caroline Clifford for her assistance in the optimization of the cell culture methods, and Mr. M Dalby for his contribution to the biomaterial testing.

REFERENCES

1. Schmidt, R. & Kulbe, K.D. (1993). Long-term cultivation of human osteoblasts. Bone Miner 20, 211-221.
2. Harris, S.A., Enger, R.J., Riggs, B.L. & Spelsberg, T.C. (1995). Development and characterization of a conditionally immortalized human fetal osteoblastic cell line. J Bone Miner Res 10, 178-186.
3. McAllister, R.M., Gardner, M.B., Greene, A.E., Bradt, C., Nichols, W.W. & Landing, B.H. (1971). Cultivation *in vitro* of cells derived from a human osteosarcoma. Cancer 27, 397-402.

4. Aubin, J.E., Heersche, J.N., Merrilees, M.J. & Sodek, J. (1982). Isolation of bone cell clones with differences in growth, hormone responses, and extracellular matrix production. J Cell Biol 92, 452-461.
5. Sudo, H., Kodama, H.A., Amagai, Y., Yamamoto, S. & Kasai, S. (1983). *In vitro* differentiation and calcification in a new clonal osteogenic cell line derived from newborn mouse calvaria. J Cell Biol 96, 191-198.
6. Owen, M. & Friedenstein, A.J. (1988). Stromal stem cells: marrow-derived osteogenic precursors. Ciba Found Symp 136, 42-60.
7. Fried, A., Benayahu, D. & Wientroub, S. (1993). Marrow stroma-derived osteogenic clonal cell lines: putative stages in osteoblastic differentiation. J Cell Physiol 155, 472-482.
8. Haynesworth, S.E., Goshima, J., Goldberg, V.M. & Caplan, A.I. (1992). Characterization of cells with osteogenic potential from human marrow. Bone 13, 81-88.
9. Friedenstein, A.J. (1976). Precursor cells of mechanocytes. Int Rev Cytol 47, 327-359.
10. Gundle, R., Joyner, C.J. & Triffitt, J.T. (1995). Human bone tissue formation in diffusion chamber culture *in vivo* by bone-derived cells and marrow stromal fibroblastic cells. Bone 16, 597-601.
11. Gallagher, J.A., Woods, D.A., Beresford, J.N., Sheard, C.E., Dillon, J.P., Russell, R.G. & Ali, S.Y. (1986). Evaluation of osteocalcin and other bone matrix proteins as markers of the osteoblastic phenotype *in vitro*. In: Cell mediated calcification and matrix vesicles, Ed. Ali, S.Y. Elsivier Science Publishers, p119-127.
12. Beresford, J.N., Gallagher, J.A., Poser, J.W. & Russell, R.G. (1984). Production of osteocalcin by human bone cells *in vitro*. Effects of 1,25(OH)2D3, 24,25(OH)2D3, parathyroid hormone, and glucocorticoids. Metab Bone Dis Relat Res 5, 229-234.
13. Jones, S.J. & Boyde, A. (1976). Experimental study of changes in osteoblastic shape induced by calcitonin and parathyroid extract in an organ culture system. Cell Tissue Res 169, 499-65.
14. Robey, P.G. & Termine, J.D. (1985). Human bone cells *in vitro*. Calcif Tissue Int 37, 453-460.
15. Aufmkolk, B. & Schwartz, E.R. (1985). Biochemical characterizations of human osteoblasts in culture. In: Normal and abnormal bone growth: Basic and clinical research, Alan R Liss Inc, p201-214.
16. Jonsson, K.B., Frost, A., Nilsson, O., Ljunghall, S. & Ljunggren, O. (1999). Three isolation techniques for primary culture of human osteoblast-like cells: a comparison. Acta Orthop Scand 70, 365-373.
17. Sell, S., Gaissmaier, C., Fritz, J., Herr, G., Esenwein, S., Kusswetter, W., Volkmann, R., Wittkowski, K.M. & Rodemann, H.P. (1998). Different behavior of human osteoblast-like cells isolated from normal and heterotopic bone In vitro. Calcif Tissue Int 62, 51-59.
18. Wong, G. & Hall, B.K. (1990). Isolation and behaviour of isolated bone forming cells. In: The osteoblast and osteocyte, Ed. Hall, B.K. Telford press, p171-191.
19. Komori, T., Yagi, H., Nomura, S., Yamaguchi, A., Sasaki, K., Deguchi, K., Shimizu, Y., Bronson, R.T., Gao, Y.H., Inada, M., Sato, M., Okamoto, R., Kitamura, Y., Yoshiki, S. & Kishimoto, T. (1997). Targeted disruption of Cbfa1 results in a complete lack of bone formation owing to maturational arrest of osteoblasts. Cell 89, 755-764.
20. Otto, F., Thornell, A.P., Crompton, T., Denzel, A., Gilmour, K.C., Rosewell, I.R., Stamp, G.W., Beddington, R.S., Mundlos, S., Olsen, B.R., Selby, P.B. & Owen, M.J. (1997). Cbfa1, a candidate gene for cleidocranial dysplasia syndrome, is essential for osteoblast differentiation and bone development. Cell 89, 765-771.
21. Ducy, P., Zhang, R., Geoffroy, V., Ridall, A.L. & Karsenty, G. (1997). Osf2/Cbfa1: a transcriptional activator of osteoblast differentiation. Cell 89, 747-754.

22. Komori, T. & Kishimoto, T. (1998). Cbfa1 in bone development. Curr Opin Genet Dev 8, 494-499.
23. Rodan, G.A. & Majeska, R.J. (1982). Phenotypic maturation of osteoblastic osteosarcoma cells in culture. Prog Clin Biol Res 110 Pt B, 249-259.
24. Ng, K.W., Partridge, N.C., Niall, M. & Martin, T.J. (1983). Epidermal growth factor receptors in clonal lines of a rat osteogenic sarcoma and in osteoblast-rich rat bone cells. Calcif Tissue Int 35, 298-303.
25. Ikeda, T., Futaesaku, Y. & Tsuchida, N. (1992). In vitro differentiation of the human osteosarcoma cell lines, HOS and KHOS. Virchows Arch B Cell Pathol Incl Mol Pathol 62, 199-206.
26. Clover, J. & Gowen, M. (1994). Are MG-63 and HOS TE85 human osteosarcoma cell lines representative models of the osteoblastic phenotype? Bone 15, 585-591.
27. Chiba, H., Sawada, N., Ono, T., Ishii, S. & Mori, M. (1993). Establishment and characterization of a simian virus 40-immortalized osteoblastic cell line from normal human bone. Jpn J Cancer Res 84, 290-297.
28. Marie, P.J. (1994). Human osteoblastic cells: a potential tool to assess the etiology of pathologic bone formation. J Bone Miner Res, 9, 1847-1850.
29. Gerstenfeld, L.C., Chipman, S.D., Glowacki, J. & Lian, J.B. (1987). Expression of differentiated function by mineralizing cultures of chicken osteoblasts. Dev Biol, 122 49-60.
30. Nefussi, J.R., Pouchelet, M., Collin, P., Sautier, J.M., Develay, G. & Forest, N. (1989). Microcinematographic and autoradiographic kinetic studies of bone cell differentiation *in vitro*: matrix formation and mineralization. Bone 10, 345-352.
31. Lomri, A. & Marie, P.J. (1988). Effect of parathyroid hormone and forskolin on cytoskeletal protein synthesis in cultured mouse osteoblastic cells. Biochim Biophys Acta 970, 333-342.
32. Majeska, R.J. & Rodan, G.A. (1982). The effect of 1,25(OH)2D3 on alkaline phosphatase in osteoblastic osteosarcoma cells. J Biol Chem 257, 3362-3365.
33. Aronow, M.A., Gerstenfeld, L.C., Owen, T.A., Tassinari, M.S., Stein, G.S. & Lian, J.B. (1990). Factors that promote progressive development of the osteoblast phenotype in cultured fetal rat calvaria cells. J Cell Physiol 143, 213-221.
34. Newman, P. & Watt, F.M. (1988). Influence of cytochalasin D-induced changes in cell shape on proteoglycan synthesis by cultured articular chondrocytes. Exp Cell Res 178, 199-210.
35. Lian, J.B. & Stein, G.S. (1992). Concepts of osteoblast growth and differentiation: basis for modulation of bone cell development and tissue formation. Crit Rev Oral Biol Med 3, 269-305.
36. Ali, S.Y. (1992). Matrix formation and mineralisation in bone. In: Bone biology and skeletal disorders, Ed. Whitehead, C.C. Carfax Publishers, Abbingdon, p19-38.
37. Riccio, V., Della, R.F., Marrone, G., Palumbo, R., Guida, G. & Oliva, A. (1994). Cultures of human embryonic osteoblasts. A new *in vitro* model for biocompatibility studies. Clin Orthop 308, 73-78.
38. Nishimoto, S.K., Stryker, W.F. & Nimni, M.E. (1987). Calcification of osteoblastlike rat osteosarcoma cells in agarose suspension cultures. Calcif Tissue Int 41, 274-280.
39. Sudo, H., Kodama, H., Amagai, Y., Itakura, Y., Yamamoto, S. & Ali, S.Y. (1986). Mineralized tissue formation by MC3T3-E1 osteogenic cells embedded in three dimensional gel matrix. In: Cell mediated calcification and matrix vesicles, Ed. Ali, S.Y. Elsivier Science Publishers, p291-296.

40. Shima, M., Seino, Y., Tanaka, H., Kurose, H., Ishida, M., Yabuuchi, H. & Kodama, H. (1988). Microcarriers facilitate mineralization in MC3T3-E1 cells. Calcif Tissue Int 43, 19-25.
41. Bellows, C.G., Aubin, J.E., Heersche, J.N. & Antosz, M.E. (1986). Mineralized bone nodules formed *in vitro* from enzymatically released rat calvaria cell populations. Calcif Tissue Int 38, 143-154.
42. Rattner, A., Sabido, O., Le, J., Vico, L., Massoubre, C., Frey, J. & Chamson, A. (2000). Mineralization and alkaline phosphatase activity in collagen lattices populated by human osteoblasts. Calcif Tissue Int 66, 35-42.
43. Casser-Bette, M., Murray, A.B., Closs, E.I., Erfle, V. & Schmidt, J. (1990). Bone formation by osteoblast-like cells in a three-dimensional cell culture. Calcif Tissue Int 46, 46-56.
44. Tenenbaum, H.C. & Heersche, J.N. (1982). Differentiation of osteoblasts and formation of mineralized bone *in vitro*. Calcif Tissue Int 34, 76-79.
45. Bellows, C.G., Aubin, J.E. & Heersche, J.N. (1987). Physiological concentrations of glucocorticoids stimulate formation of bone nodules from isolated rat calvaria cells *in vitro*. Endocrinology 121, 1985-1992.
46. Koshihara, Y., Kawamura, M., Endo, S., Tsutsumi, C., Kodama, H., Oda, H. & Higaki, S. (1989). Establishment of human osteoblastic cells derived from periosteum in culture. In Vitro Cell Dev Biol 25, 37-43.
47. Matsuyama, T., Lau, K.H. & Wergedal, J.E. (1990). Monolayer cultures of normal human bone cells contain multiple subpopulations of alkaline phosphatase positive cells. Calcif Tissue Int 47, 276-283.
48. Fedarko, N.S., Bianco, P., Robey, G.P. & Termine, J.D. (1989). Alkaline phosphatase activity in human bone cell cultures is cell cycle dependent. J Bone Miner Res 4, 121
49. Stringa, E., Filanti, C., Giunciuglio, D., Albini, A. & Manduca, P. (1995). Osteoblastic cells from rat long bone. I. Characterization of their differentiation in culture. Bone 16, 663-670.
50. Stutte, H.J. (1967). [Hexazotated triamino-tritolyl-methane chloride (neofuchsin) as developer in enzyme histochemistry]. Histochemie, 8, 327-331.
51. Stein, G.S., Lian, J.B. & Owen, T.A. (1990). Relationship of cell growth to the regulation of tissue-specific gene expression during osteoblast differentiation. FASEB J 4, 3111-3123.
52. Canalis, E., McCarthy, T. & Centrella, M. (1988). Growth factors and the regulation of bone remodeling. J Clin Invest 81, 277-281.
53. Mohan, S. & Baylink, D.J. (1991). Bone growth factors. Clin Orthop 263, 30-48.
54. Stracke, H., Schulz, A., Moeller, D., Rossol, S. & Schatz, H. (1984). Effect of growth hormone on osteoblasts and demonstration of somatomedin-C/IGF I in bone organ culture. Acta Endocrinol (Copenh) 107, 16-24.
55. Ernst, M. & Froesch, E.R. (1988). Growth hormone dependent stimulation of osteoblast-like cells in serum-free cultures via local synthesis of insulin-like growth factor I. Biochem Biophys Res Commun 151, 142-147.
56. Slootweg, M.C., van Buul-Offers, S.C., Herrmann-Erlee, M.P., van der Meer, J.M. & Duursma, S.A. (1988). Growth hormone is mitogenic for fetal mouse osteoblasts but not for undifferentiated bone cells. J Endocrinol 116, R11-R13
57. Maor, G., Hochberg, Z., von der, M., Heinegard, D. & Silbermann, M. (1989). Human growth hormone enhances chondrogenesis and osteogenesis in a tissue culture system of chondroprogenitor cells. Endocrinology 125, 1239-1245.
58. Scheven, B.A., Hamilton, N.J., Fakkeldij, T.M. & Duursma, S.A. (1991). Effects of recombinant human insulin-like growth factor I and II (IGF-I/-II) and growth hormone

(GH) on the growth of normal adult human osteoblast-like cells and human osteogenic sarcoma cells. Growth Regul 1, 160-167.

59. Kassem, M., Blum, W., Ristelli, J., Mosekilde, L. & Eriksen, E.F. (1993). Growth hormone stimulates proliferation and differentiation of normal human osteoblast-like cells *in vitro*. Calcif Tissue Int 52, 222-226.
60. Chenu, C., Valentin-Opran, A., Chavassieux, P., Saez, S., Meunier, P.J. & Delmas, P.D. (1990). Insulin like growth factor I hormonal regulation by growth hormone and by 1,25(OH)2D3 and activity on human osteoblast-like cells in short-term cultures. Bone 11, 81-86.
61. Di-Silvio, L. A novel application of two biomaterials for the delivery of Growth Hormone and its effect on osteoblasts. 1995. PhD Thesis , University of London.
62. Brixen, K., Nielsen, H.K., Mosekilde, L. & Flyvbjerg, A. (1990). A short course of recombinant human growth hormone treatment stimulates osteoblasts and activates bone remodeling in normal human volunteers. J Bone Miner Res 5, 609-618.
63. Slootweg, M.C., van Buul-Offers, S.C., Herrmann-Erlee, M.P. & Duursma, S.A. (1988). Direct stimulatory effect of growth hormone on DNA synthesis of fetal chicken osteoblasts in culture. Acta Endocrinol (Copenh) 118, 294-300.
64. Morel, G., Chavassieux, P., Barenton, B., Dubois, P.M., Meunier, P.J. & Boivin, G. (1993). Evidence for a direct effect of growth hormone on osteoblasts. Cell Tissue Res 273, 279-286.
65. Schmid, C., Zapf, J. & Froesch, E.R. (1989). Production of carrier proteins for insulin-like growth factors (IGFs) by rat osteoblastic cells. Regulation by IGF I and cortisol. FEBS Lett 244, 328-332.
66. Katzburg, S., Lieberherr, M., Ornoy, A., Klein, B.Y., Hendel, D. & Somjen, D. (1999). Isolation and hormonal responsiveness of primary cultures of human bone-derived cells: gender and age differences. Bone 25, 667-673.
67. Dalby, M.J., Di-Silvio, L., Harper, E.J. & Bonfield, W. (1999). In vitro evaluation of a new cement polymethylmethacrylate cement reinforced with hydroxyapatite. J of Mater Sci: Mats in Med 10, 793-796.
68. Di-Silvio, L., Huang, J. & Bonfield, W. (1999). The *in vitro* behaviour of human osteoblast-like cells on hydroxyapatite and Bioglass-reinforced polyethylene composites. In: Ceramics, Cells and Tissues: Ceramic-polymer composites., Ed. Ravaglioli, A. and Krajewski, A. IRTEC-CNR, Faenza. p201-206.
69. Huang, J., Hing, K. A., Best, S. M., Lee, D. A., Di-Silvio, L., and Bonfield, W. (1999). A comparative study of various assays for *in vitro* biocompatibility assessment of experimental zirconia. In: Bioceramics:Proceedings of the 12th international symposium on ceramics in medicine. Eds. Ohgushi, H., Yoshikawa, T., and Hastings, G. W., World Scientific Pub Co., p199-202.
70. Williams, D.F. (1987). Theoritical and practical aspects of testing potential biomaterials *in vitro*. J of Mater Sci: Mats in Med 1, 9-13.
71. Deb, S., Di-Silvio, L., Vazquez, B. & San Roman, J. (1999). Water absorption characteristics and cytotoxic and biological evaluation of bone cements formulated with a novel activator. J Biomed Mater Res 48, 719-725.
72. Hing, K A., Di-Silvio, L., Gibson, IR., Ohtsuki, C., Jha, LJ., Best, SM., and Bonfield, W. (1997). Effect of Fluoride substitution on the biocompatibility of Hydroxyapaptie. In: Proceedings of the 10th international symposium on ceramics in medicine. Eds. Sedel, L. and Rey, C., Elsevier Scence Limited. Paris, p19-22.
73. Hing, K. A., Gibson, I. R., Di-Silvio, L., Best, S. M., and Bonfield, W. (1998). Effect of variation in Ca:P ration on cellular respomse of primary human osteoblast-like cells to hydroxyapatite-based ceramics. In: Proceedings of the 11th international symposium on

ceramics in medicine. Eds. LeGeros, R. Z. and leGeros, J. P. World Scientific Publishing Co. New York, p293-296.
74. Bonfield, W., Doyle, C. & Tanner, K.E. (1986). In vivo evaluation of hydroxypapatite reinforced polyethylene composites. In: Biological and biomechanical performance of biomaterials, Eds. Christel, P., Meunier, C.A. & Lee, A. Elsivier, Amstradam. p153-159.
75. Bonfield, W. (1988). Hydroxyapaptite reinforced polyethylene as an analogous materials for bone replacement. In: Bioceramics: Materials characteristics versus *in vivo* behaviour, Eds. Ducheyne, P. and Lemmons, JE. Annals of New York academy of sciences, New York. p173-175.
76. Huang, J., Di-Silvio, L., Wang, M., Tanner, K.E. & Bonfield, W. (1997). In vitro mechanical assessment of hydroxyapatite reinforced polyethylene composite. J of Mater Sci: Mats in Med 8, 809-813.
77. Huang, J., Di-Silvio, L., Wang, M., Tanner, K. E., and Bonfield, W. (1997). In vitro assesment of hydroxyapaptite and Bioglass reinforced polyethylene composites. In: Bioceramics; Proceedings of the 10 th international symposium on ceramics in medicine. Eds. Sedel, L. and Rey, C. Paris, Elsevier Science Limited. p519-522.
78. Di-Silvio, L., Dalby, M.J. & Bonfield, W. (1998). In vitro response of osteoblasts to hydroxyapatite-reinforced polyethylene composites. J of Mater Sci: Mats in Med 9, 845-848.
79. Dalby, M. J., Di-Silvio, L., and Bonfield, W. (2000). Osteoblast response to Hapex surface topography. In: The sixth world biomaterials congress transactions. Hawaii, USA. p673.
80. Di-Silvio, L., Jayakumar, P., Hing, K. A., Best, S. M., and Bonfield, W. (2000). Growth factor containing macroporous hydroxyapatite for bone tissue engineering. In: The sixth world biomaterials congress transactions. Hawaii, USA. p440.
81. Evans, C.E., Ng, K., Allen, J. & Gallimore, P. (1995). Modulation of cell phenotype in human osteoblast-like cells by the simian virus 40. J Orthop Res 13, 317-324.
82. Grigoriadis, A.E., Petkovich, P.M., Ber, R., Aubin, J.E. & Heersche, J.N. (1985). Subclone heterogeneity in a clonally-derived osteoblast-like cell line. Bone 6, 249-256.

Human Cell Culture

1. J.R.W. Masters and B. Palsson (eds.): *Human Cell Culture. Vol. I.* 1998
ISBN 0-7923-5143-6

2. J.R.W. Masters and B. Palsson (eds.): *Human Cell Culture, Part 2. Vol. II.* Cancer Cell Lines. 1999
ISBN 0-7923-5878-3

3. J.R.W. Masters and B.O. Palsson (eds.): *Human Cell Culture, Part 3.* Cancer Continuous Cell Lines: Leukemias and Lymphomas. 2000
ISBN 0-7923-6225-X

4. M.R. Koller, B.O. Palsson and J.R.W. Masters (eds.): *Human Cell Culture, Volume IV.* Primary Hematopoietic Cells. 1999
ISBN 0-7923-5821-X

5. M.R. Koller, B.O. Palsson and J.R.W. Masters (eds.): *Human Cell Culture, Volume V.* Primary Mesenchymal Cells. 2001
ISBN 0-7923-6761-8

KLUWER LAW INTERNATIONAL – THE HAGUE / LONDON / BOSTON

MIX
Papier aus verantwortungsvollen Quellen
Paper from responsible sources
FSC® C105338

If you have any concerns about our products,
you can contact us on
ProductSafety@springernature.com

In case Publisher is established outside the EU,
the EU authorized representative is:
Springer Nature Customer Service Center GmbH
Europaplatz 3, 69115 Heidelberg, Germany

Printed by Libri Plureos GmbH
in Hamburg, Germany